THE DOUBLE HELIX

A Personal Account of the Discovery of the Structure of DNA

by James D. Watson

A SIGNET BOOK

Published by
THE NEW AMERICAN LIBRARY

Library of Congress Catalog Card Number: 68-16217

This is an authorized reprint of a hardcover edition published by Atheneum. The hardcover edition was published simultaneously in Canada by McClelland and Stewart Ltd. Portions of this book were first published in *The Atlantic Monthly*.

SIGNET BOOKS are published by
The New American Library, Inc.,
1301 Avenue of the Americas, New York, New York 10019

FIRST PRINTING, FEBRUARY, 1969

PRINTED IN THE UNITED STATES OF AMERICA

"IT IS A STRANGE MODEL AND EMBODIES SEVERAL UNUSUAL FEATURES. HOWEVER, SINCE DNA IS AN UNUSUAL SUBSTANCE, WE ARE NOT HESITANT IN BEING BOLD."

Thus Jim Watson, aged twenty-four, wrote from Cambridge to a friend in the States, one month before the public announcement of a discovery many scientists now call the most significant since Mendel's.

***The Double Helix* is the story of that startling breakthrough. It is about science. Scientists. Scientific politics. Girls. Wine. Movies. Cambridge and London and Paris. Teamwork. Rivalry. Genius. Stupidity. And virtually everything else that make life in the lab and out something less than perfect and a great deal more than dull.**

"Perfectly fascinating and delightful, . . . engaging directness and awesome candor"

—Carl Kaysen,
Director, Institute of
Advanced Studies, Princeton

"He has described admirably how it feels to have that frightening and beautiful experience of making a great scientific discovery"

—Richard Feynman,
Nobel Prize Winner, 1965

"Human, funny, cross-grained, emotional . . . an unexpected breeze of fresh air from a region men have thought of as stuffy and incommunicado, the world of pure science"—*Publishers' Weekly*

For Naomi Mitchison

Foreword by Sir Lawrence Bragg

THIS ACCOUNT of the events which led to the solution of the structure of DNA, the fundamental genetical material, is unique in several ways. I was much pleased when Watson asked me to write the foreword.

There is in the first place its scientific interest. The discovery of the structure by Crick and Watson, with all its biological implications, has been one of the major scientific events of this century. The number of researches which it has inspired is amazing; it has caused an explosion in biochemistry which has transformed the science. I have been amongst those who have pressed the author to write his recollections while they are still fresh in his mind, knowing how important they would be as a contribution to the history of science. The result has exceeded expectation. The latter chapters, in which the birth of the new idea is described so vividly, are drama of the highest order; the tension mounts and mounts towards the final climax. I do not know of any other instance where one is able to share so intimately in the researcher's struggles and doubts and final triumph.

Then again, the story is a poignant example of a dilemma which may confront an investigator. He knows that a colleague has been working for years on a problem and has accumulated a mass of hard-won evidence, which has not yet been published because it is anticipated that success is just around the corner. He has seen this evidence and has good reason to believe that a method of attack which he can envisage, perhaps merely a new point of view, will lead straight to the solution. An offer of collaboration at such a stage might well be regarded as a trespass. Should he go ahead on his own? It is not easy to be sure whether the crucial new idea is really one's own or has been unconsciously assimilated in talks with others. The realization of this difficulty has led to the establishment of a somewhat vague code amongst scientists which recognizes a claim in a line of research staked out by a colleague—up to a certain point. When competition comes from more than one quarter, there is no need to hold back. This dilemma comes out clearly in the DNA story. It is a source of deep satisfaction to all intimately concerned that, in the award of the Nobel Prize in 1962, due recognition was given to the long, patient investigation by Wilkins at King's College (London) as well as to the brilliant and rapid final solution by Crick and Watson at Cambridge.

Finally, there is the human interest of the story—the impression made by Europe and by England in particular upon a young man from the States. He writes with a Pepys-like frankness. Those who figure in the book must read it in a very forgiving spirit. One must remember that his book is not a history, but an autobiographical contribution to the history which will some day be written. As the author himself says, the book is a record of impressions rather than historical facts. The issues were often more complex, and the motives of those who had to deal with them were less tortuous, than he realized at the time. On the other hand, one must admit that his intuitive understanding of human frailty often strikes home.

The author has shown the manuscript to some of us who were involved in the story, and we have suggested corrections of historical fact here and there, but personally I have felt reluctant to alter too much because the freshness and directness with which impressions have been recorded is an essential part of the interest of this book.

W. L. B.

Preface

HERE I relate my version of how the structure of DNA was discovered. In doing so I have tried to catch the atmosphere of the early postwar years in England, where most of the important events occurred. As I hope this book will show, science seldom proceeds in the straightforward logical manner imagined by outsiders. Instead, its steps forward (and sometimes backward) are often very human events in which personalities and cultural traditions play major roles. To this end I have attempted to re-create my first impressions of the relevant events and personalities rather than present an assessment which takes into account the many facts I have learned since the structure was found. Although the latter approach might be more objective, it would fail to convey the spirit of an adventure characterized both by youthful arrogance and by the belief that the truth, once found, would be simple as well as pretty. Thus many of the comments may seem one-sided and unfair, but this is often the case in the incomplete and hurried way in which human beings frequently decide to like or dislike a new idea or acquaintance. In any event, this account represents the way I saw things then, in 1951–1953: the ideas, the people, and myself.

I am aware that the other participants in this story would tell parts of it in other ways, sometimes because their memory of what happened differs from mine and, perhaps in even more cases, because no two people ever see the same events in exactly the same light. In this sense, no one will ever be able to write a definitive history of how the structure was established. Nonetheless, I feel the story should be told, partly because many of my scientific friends have expressed curiosity about how the double helix was found, and to them an incomplete version is better than none. But even more important, I believe, there remains general ignorance about how science is "done." That is not to say that all science is done in the manner described here. This is far from the case, for styles of scientific research vary almost as much as human personalities. On the other hand, I do not believe that the way DNA came out constitutes an odd exception to a scientific world complicated by the contradictory pulls of ambition and the sense of fair play.

The thought that I should write this book has been with me almost from the moment the double helix was found. Thus my memory of many of the significant events is much more complete than that of most other episodes in my life. I also have made extensive use of letters written at virtually weekly intervals to my parents. These were especially helpful in exactly dating a number of the incidents. Equally important have been the valuable comments by various friends who kindly read earlier versions and gave in some instances quite detailed accounts of incidents that I had referred to in less complete form. To be sure, there are cases where my recollections differ from theirs, and so this book must be regarded as my view of the matter.

Some of the earlier chapters were written in the homes of Albert Szent-Györgyi, John A. Wheeler, and John Cairns, and I wish to thank them for quiet rooms with tables overlooking the ocean. The later chapters were written with the help of a Guggenheim Fellowship, which allowed me to return briefly to the other Cambridge and the kind hospitality of the Provost and Fellows of King's College.

As far as possible I have included photographs taken at the time the story occurred, and in particular I want to thank Herbert Gutfreund, Peter Pauling, Hugh Huxley, and Gunther Stent for sending me some of their snapshots. For editorial assistance I'm much indebted to Libby Aldrich for the quick, perceptive remarks expected from our best Radcliffe students and to Joyce Lebowitz both for keeping me from completely misusing the English language and for innumerable comments about what a good book must do. Finally, I wish to express thanks for the immense help Thomas J. Wilson has given me from the time he saw the first draft. Without his wise, warm, and sensible advice, the appearance of this book, in what I hope is the right form, might never have occurred.

J. D. W.

Harvard University
Cambridge, Massachusetts
November 1967

Diagrams

In the summer of 1955, I arranged to join some friends who were going into the Alps. Alfred Tissieres, then a Fellow at King's, had said he would get me to the top of the Rothorn, and even though I panic at voids this did not seem to be the time to be a coward. So after getting in shape by letting a guide lead me up the Allinin, I took the two-hour postal-bus trip to Zinal, hoping that the driver was not carsick as he lurched the bus around the narrow road twisting above the falling rock slopes. Then I saw Alfred standing in front of the hotel, talking with a long-mustached Trinity don who had been in India during the war.

Since Alfred was still out of training, we decided to spend the afternoon walking up to a small restaurant which lay at the base of the huge glacier falling down off the Obergabelhorn and over which we were to walk the next day. We were only a few minutes out of sight of the hotel when we saw a party coming down upon us, and I quickly recognized one of the climbers. He was Willy Seeds, a scientist who several years before had worked at King's College, London, with Maurice Wilkins on the optical properties of DNA fibers. Willy soon spotted me, slowed down, and momentarily gave the impression that he might remove his rucksack and chat for a while. But all he said was, "How's Honest Jim?" and quickly increasing his pace was soon below me on the path.

Later as I trudged upward, I thought again about our earlier meetings in London. Then DNA was still a mystery, up for grabs, and no one was sure who would get it and whether he would deserve it if it proved as exciting as we semisecretly believed. But now the race was over and, as one of the winners, I knew the tale was not simple and certainly not as the newspapers reported. Chiefly it was a matter of five people: Maurice Wilkins, Rosalind Franklin, Linus Pauling, Francis Crick, and me. And as Francis was the dominant force in shaping my part, I will start the story with him.

1

I HAVE never seen Francis Crick in a modest mood. Perhaps in other company he is that way, but I have never had reason so to judge him. It has nothing to do with his present fame. Already he is much talked about, usually with reverence, and someday he may be considered in the category of Rutherford or Bohr. But this was not true when, in the fall of 1951, I came to the Cavendish Laboratory of Cambridge University to join a small group of physicists and chemists working on the three-dimensional structures of proteins. At that time he was thirty-five, yet almost totally unknown. Although some of his closest colleagues realized the value of his quick, penetrating mind and frequently sought his advice, he was often not appreciated, and most people thought he talked too much.

Leading the unit to which Francis belonged was Max Perutz, an Austrian-born chemist who came to England in 1936. He had been collecting X-ray diffraction data from hemoglobin crystals for over ten years and was just beginning to get somewhere. Helping him was Sir Lawrence Bragg, the director of the Cavendish. For almost forty years Bragg, a Nobel Prize winner and one of the founders of crystallography, had been watching X-ray diffraction methods solve structures of ever-increasing difficulty.* The more complex the molecule, the happier Bragg became when a new method allowed its elucidation. Thus in the immediate postwar years he was especially keen about the possibility of solving the structures of proteins, the most complicated of all molecules. Often, when administrative duties permitted, he visited Perutz' office to discuss recently accumulated X-ray data. Then he would return home to see if he could interpret them.

* For a clear description of X-ray diffraction technique, see John Kendrew, *The Thread of Life: An Introduction to Molecular Biology* (Cambridge: Harvard University Press, 1966), p. 14.

Somewhere between Bragg the theorist and Perutz the experimentalist was Francis, who occasionally did experiments but more often was immersed in the theories for solving protein structures. Often he came up with something novel, would become enormously excited, and immediately tell it to anyone who would listen. A day or so later he would often realize that his theory did not work and return to experiments, until boredom generated a new attack on theory.

There was much drama connected with these ideas. They did a great deal to liven up the atmosphere of the lab, where experiments usually lasted several months to years. This came partly from the volume of Crick's voice: he talked louder and faster than anyone else and, when he laughed, his location within the Cavendish was obvious. Almost everyone enjoyed these manic moments, especially when we had the time to listen attentively and to tell him bluntly when we lost the train of his argument. But there was one notable exception. Conversations with Crick frequently upset Sir Lawrence Bragg, and the sound of his voice was often sufficient to make Bragg move to a safer room. Only infrequently would he come to tea in the Cavendish, since it meant enduring Crick's booming over the tea room. Even then Bragg was not completely safe. On two occasions the corridor outside his office was flooded with water pouring out of a laboratory in which Crick was working. Francis, with his interest in theory, had neglected to fasten securely the rubber tubing around his suction pump.

At the time of my arrival, Francis' theories spread far beyond the confines of protein crystallography. Anything important would attract him, and he frequently visited other labs to see which new experiments had been done. Though he was generally polite and considerate of colleagues who did not realize the real meaning of their latest experiments, he would never hide this fact from them. Almost immediately he would suggest a rash of new experiments that should confirm his interpretation. Moreover, he would not refrain from subsequently telling all who would listen how his clever new idea might set science ahead.

As a result, there existed an unspoken yet real fear of Crick, especially among his contemporaries who had yet to establish their reputations. The quick manner in which he seized their facts and tried to reduce them to coherent patterns frequently made his friends' stomachs sink with the apprehension that, all too often in the near future, he would succeed, and expose to the world the fuzziness of minds hidden from direct view by the considerate, well-spoken manners of the Cambridge colleges.

Though he had dining rights for one meal a week at Caius College, he was not yet a fellow of any college. Partly this was his own choice. Clearly he did not want to be burdened by the unnecessary sight of undergraduate tutees. Also a factor was his laugh, against which many dons would most certainly rebel if subjected to its shattering bang more than once a week. I am sure this occasionally bothered Francis, even though he obviously knew that most High Table life is dominated by pedantic, middle-aged men incapable of either amusing or educating him in anything worthwhile. There always existed King's College, opulently nonconformist and clearly capable of absorbing him without any loss of his or its character. But despite much effort on the part of his friends, who knew he was a delightful dinner companion, they were never able to hide the fact that a stray remark over sherry might bring Francis smack into your life.

2

BEFORE my arrival in Cambridge, Francis only occasionally thought about deoxyribonucleic acid (DNA) and its role in heredity. This was not because he thought it uninteresting. Quite the contrary. A major factor in his leaving physics and developing an interest in biology had been the reading in 1946 of *What Is Life?* by the noted theoretical physicist Erwin Schrödinger. This book very elegantly propounded the belief that genes were the key components of living cells and that, to understand what life is, we must know how genes act. When Schrödinger wrote his book (1944), there was general acceptance that genes were special types of protein molecules. But almost at this same time the bacteriologist O. T. Avery was carrying out experiments at the Rockefeller Institute in New York which showed that hereditary traits could be transmitted from one bacterial cell to another by purified DNA molecules.

Given the fact that DNA was known to occur in the chromosomes of all cells, Avery's experiments strongly suggested that future experiments would show that all genes were composed of DNA. If true, this meant to Francis that proteins would not be the Rosetta Stone for unraveling the true secret of life. Instead, DNA would have to provide the key to enable us to find out how the genes determined, among other characteristics, the color of our hair, our eyes, most likely our comparative intelligence, and maybe even our potential to amuse others.

Of course there were scientists who thought the evidence favoring DNA was inconclusive and preferred to believe that genes were protein molecules. Francis, however, did not worry about these skeptics. Many were cantankerous fools who unfailingly backed the wrong horses. One could not be a successful scientist without realizing that, in contrast to the popular conception supported by newspapers and mothers of scientists, a goodly number of scientists are not only narrow-minded and dull, but also just stupid.

Francis, nonetheless, was not then prepared to jump into the DNA world. Its basic importance did not seem sufficient cause by itself to lead him out of the protein field which he had worked in only two years and was just beginning to master intellectually. In addition, his colleagues at the Cavendish were only marginally interested in the nucleic acids, and even in the best of financial circumstances it would take two or three years to set up a new research group primarily devoted to using X rays to look at the DNA structure.

Moreover, such a decision would create an awkward personal situation. At this time molecular work on DNA in England was, for all practical purposes, the personal property of Maurice Wilkins, a bachelor who worked in London at King's College* Like Francis, Maurice had been a physicist and also used X-ray diffraction as his principal tool of research. It would have looked very bad if Francis had jumped in on a problem that Maurice had worked over for several years. The matter was even worse because the two, almost equal in age, knew each other and, before Francis remarried, had frequently met for lunch or dinner to talk about science.

It would have been much easier if they had been living in different countries. The combination of England's coziness—all the important people, if not related by marriage, seemed to know one another—plus the English sense of fair play would not allow Francis to move in on Maurice's problem. In France, where fair play obviously did not exist, these problems would not have arisen. The States also would not have permitted such a situation to develop. One would not expect someone at Berkeley to ignore a first-rate problem merely because someone at Cal Tech had started first. In England, however, it simply would not look right.

Even worse, Maurice continually frustrated Francis by never seeming enthusiastic enough about DNA. He appeared to enjoy slowly understating important arguments. It was not a question of intelligence or common sense.

* A division of the University of London, not to be confused with King's College, Cambridge.

Maurice clearly had both; witness his seizing DNA before almost everyone else. It was that Francis felt he could never get the message over to Maurice that you did not move cautiously when you were holding dynamite like DNA. Moreover, it was increasingly difficult to take Maurice's mind off his assistant, Rosalind Franklin.

Not that he was at all in love with Rosy, as we called her from a distance. Just the opposite—almost from the moment she arrived in Maurice's lab, they began to upset each other. Maurice, a beginner in X-ray diffraction work, wanted some professional help and hoped that Rosy, a trained crystallographer, could speed up his research. Rosy, however, did not see the situation this way. She claimed that she had been given DNA for her own problem and would not think of herself as Maurice's assistant.

I suspect that in the beginning Maurice hoped that Rosy would calm down. Yet mere inspection suggested that she would not easily bend. By choice she did not emphasize her feminine qualities. Though her features were strong, she was not unattractive and might have been quite stunning had she taken even a mild interest in clothes. This she did not. There was never lipstick to contrast with her straight black hair, while at the age of thirty-one her dresses showed all the imagination of English blue-stocking adolescents. So it was quite easy to imagine her the product of an unsatisfied mother who unduly stressed the desirability of professional careers that could save bright girls from marriages to dull men. But this was not the case. Her dedicated, austere life could not be thus explained—she was the daughter of a solidly comfortable, erudite banking family.

Clearly Rosy had to go or be put in her place. The former was obviously preferable because, given her belligerent moods, it would be very difficult for Maurice to maintain a dominant position that would allow him to think unhindered about DNA. Not that at times he didn't see some reason for her complaints—King's had two combination rooms, one for men, the other for women, certainly a thing of the past. But he was not responsible, and it was no pleasure to bear the cross for the added barb that the women's combination room remained dingily pokey whereas money had been spent to make life agreeable for him and his friends when they had their morning coffee.

Unfortunately, Maurice could not see any decent way to give Rosy the boot. To start with, she had been given to think that she had a position for several years. Also, there was no denying she had a good brain. If she could only keep her emotions under control, there would be a good chance that she could really help him. But merely wishing for relations to improve was taking something of a gamble, for Cal Tech's fabulous chemist Linus Pauling was not subject to the confines of British fair play. Sooner or later Linus, who had just turned fifty, was bound to try for the most important of all scientific prizes. There was no doubt that he was interested. Our first principles told us that Pauling could not be the greatest of all chemists without realizing that DNA was the most golden of all molecules. Moreover, there was definite proof. Maurice had received a letter from Linus asking for a copy of the crystalline DNA X-ray photographs. After some hesitation he wrote back saying that he wanted to look more closely at the data before releasing the pictures.

All this was most unsettling to Maurice. He had not escaped into biology only to find it personally as objectionable as physics, with its atomic consequences. The combination of both Linus and Francis breathing down his neck often made it very difficult to sleep. But at least Pauling was six thousand miles away, and even Francis was separated by a two-hour rail journey. The real problem, then, was Rosy. The thought could not be avoided that the best home for a feminist was in another person's lab.

3

IT WAS Wilkins who had first excited me about X-ray work on DNA. This happened at Naples when a small scientific meeting was held on the structures of the large molecules found in living cells. Then it was the spring of 1951, before I knew of Francis Crick's existence. Already I was much involved with DNA, since I was in Europe on a postdoctoral fellowship to learn its biochemistry. My interest in DNA had grown out of a desire, first picked up while a senior in college, to learn what the gene was. Later, in graduate school at Indiana University, it was my hope that the gene might be solved without my learning any chemistry. This wish partially arose from laziness since, as an undergraduate at the University of Chicago, I was principally interested in birds and managed to avoid taking any chemistry or physics courses which looked of even medium difficulty. Briefly the Indiana biochemists encouraged me to learn organic chemistry, but after I used a bunsen burner to warm up some benzene, I was relieved from further true chemistry. It was safer to turn out an uneducated Ph.D. than to risk another explosion.

So I was not faced with the prospect of absorbing chemistry until I went to Copenhagen to do my postdoctoral research with the biochemist Herman Kalckar. Journeying abroad initially appeared the perfect solution to the complete lack of chemical facts in my head, a condition at times encouraged by my Ph.D. supervisor, the Italian-trained microbiologist Salvador Luria. He positively abhorred most chemists, especially the competitive variety out of the jungles of New York City. Kalckar, however, was obviously cultivated, and Luria hoped that in his civilized, continental company I would learn the necessary tools to do chemical research, without needing to react against the profit-oriented organic chemists.

Then Lurai's experiments largely dealt with the multiplication of bacterial viruses (bacteriophages, or phages for short). For some years the suspicion had existed among the more inspired geneticists that viruses were a form of naked genes. If so, the best way to find out what a gene was and how it duplicated was to study the properties of viruses. Thus, as the simplest viruses were the phages, there had sprung up between 1940 and 1950 a growing number of scientists (the phage group) who studied phages with the hope that they would eventually learn how the genes controlled cellular heredity. Leading this group were Luria and his German-born friend, the theoretical physicist Max Delbrück, then a professor at Cal Tech. While Delbrück kept hoping that purely genetic tricks could solve the problem, Luria more often wondered whether the real answer would come only after the chemical structure of a virus (gene) had been cracked open. Deep down he knew that it is impossible to describe the behavior of something when you don't know what it is. Thus, knowing he could never bring himself to learn chemistry, Luria felt the wisest course was to send me, his first serious student, to a chemist.

He had no difficulty deciding between a protein chemist and a nucleic-acid chemist. Though only about one half the mass of a bacterial virus was DNA (the other half being protein), Avery's experiment made it smell like the essential genetic material. So working out DNA's chemical structure might be the essential step in learning how genes duplicated. Nonetheless, in contrast to the proteins, the solid chemical facts known about DNA were meager. Only a few chemists worked with it and, except for the fact that nucleic acids were very large molecules built up from smaller building blocks, the nucleotides, there was almost nothing chemical that the geneticist could grasp at. Moreover, the chemists who did work on DNA were almost always organic chemists with no interest in genetics. Kalckar was a bright exception. In the summer of 1945 he had come to the lab at Cold Spring Harbor, New York, to take Delbrück's course on bacterial viruses. Thus both Luria and Delbrück hoped the Copenhagen lab would be the place where the combined techniques of chemistry and genetics might eventually yield real biological dividends.

Their plan, however, was a complete flop. Herman did not stimulate me in the slightest. I found myself just as indifferent to nucleic-acid chemistry in his lab as I had been in the States. This was partly because I could not see how the type of problem on which he was then working (the metabolism of nucleotides) would lead to anything of immediate interest to genetics. There was also the fact that, though Herman was obviously civilized, it was impossible to understand him.

I was able, however, to follow the English of Herman's close friend Ole Maaløe. Ole had just returned from the States (Cal Tech), where he had become very excited about the same phages on which I had worked for my degree. Upon his return he gave up his previous research problem and was devoting full time to phage. Then he was the only Dane working with phage and so was quite pleased that I and Gunther Stent, a phage worker from Delbrück's lab, had come to do research with Herman. Soon Gunther and I found ourselves going regularly to visit Ole's lab, located several miles from Herman's, and within several weeks we were both actively doing experiments with Ole.

At first I occasionally felt ill at ease doing conventional phage work with Ole, since my fellowship was explicitly awarded to enable me to learn biochemistry with Herman; in a strictly literal sense I was violating its terms. Moreover, less than three months after my arrival in Copenhagen I was asked to propose plans for the following year. This was no simple matter, for I had no plans. The only safe course was to ask for funds to spend another year with Herman. It would have been risky to say that I could not make myself enjoy biochemistry. Furthermore, I could see no reason why they should not permit me to change my plans after the renewal was granted. I thus wrote to Washington saying that I wished to remain in the stimulating environment of Copenhagen. As expected, my fellowship was then renewed. It made sense to let Kalckar (whom several of the fellowship electors knew personally) train another biochemist.

There was also the question of Herman's feelings. Perhaps he minded the fact that I was only seldom around. True, he appeared very vague about most things and might not yet have really noticed. Fortunately, however, these fears never had time to develop seriously. Through a completely unanticipated event my moral conscience became clear. One day early in December, I cycled over to Herman's lab expecting another charming yet totally incomprehensible conversation. This time, however, I found Herman could be understood. He had something important to let out: his marriage was over, and he hoped to obtain a divorce. This fact was soon no secret—everyone else in the lab was also told. Within a few days it became apparent that Herman's mind was not going to concentrate on science for some time, for perhaps as long as I would remain in Copenhagen. So the fact that he did not have to teach me nucleic-acid biochemistry was obviously a godsend. I could cycle each day over to Ole's lab, knowing it was clearly better to deceive the fellowship electors about where I was working than to force Herman to talk about biochemistry.

At times, moreover, I was quite pleased with my current experiments on bacterial viruses. Within three months Ole and I had finished a set of experiments on the fate of a bacterial-virus particle when it multiplies inside a bacterium to form several hundred new virus particles. There were enough data for a respectable publication and, using ordinary standards, I knew I could stop work for the rest of the year without being judged unproductive. On the other hand, it was equally obvious that I had not done anything which was going to tell us what a gene was or how it reproduced. And unless I became a chemist, I could not see how I would.

I thus welcomed Herman's suggestion that I go that spring to the Zoological Station at Naples, where he had decided to spend the months of April and May. A trip to Naples made great sense. There was no point in doing nothing in Copenhagen, where spring does not exist. On the other hand, the sun of Naples might be conducive to learning something about the biochemistry of the embryonic development of marine animals. It might also be a place where I could quietly read genetics. And when I was tired of it, I might conceivably pick up a biochemistry text. Without any hesitation I wrote to the States requesting permission to accompany Herman to Naples. A cheerful affirmative letter wishing me a pleasant journey came by return post from Washington. Moreover, it enclosed a $200 check for travel expenses. It made me feel slightly dishonest as I set off for the sun.

4

MAURICE WILKINS also had not come to Naples for serious science. The trip from London was an unexpected gift from his boss, Professor J. T. Randall. Originally Randall had been scheduled to come to the meeting on macromolecules and give a paper about the work going on in his new biophysics lab. Finding himself overcommitted, he had decided to send Maurice instead. If no one went, it would look bad for his King's College lab. Lots of scarce Treasury money had to be committed to set up his biophysics show, and suspicions existed that this was money down the drain.

No one was expected to prepare an elaborate talk for Italian meetings like this one. Such gatherings routinely brought together a small number of invited guests who did not understand Italian and a large number of Italians, almost none of whom understood rapidly spoken English, the only language common to the visitors. The high point of each meeting was the day-long excursion to some scenic house or temple. Thus there was seldom chance for anything but banal remarks.

By the time Maurice arrived I was noticeably restless and impatient to return north. Herman had completely misled me. For the first six weeks in Naples I was constantly cold. The official temperature is often much less relevant than the absence of central heating. Neither the Zoological Station nor my decaying room atop a six-story nineteenth-century house had any heat. If I had even the slightest interest in marine animals, I would have done experiments. Moving about doing experiments is much warmer than sitting in the library with one's feet on a table. At times I stood about nervously while Herman went through the motions of a biochemist, and on several days I even understood what he said. It made no difference, however, whether or not I followed the argument.

Genes were never at the center, or even at the periphery, of his thoughts.

Most of my time I spent walking the streets or reading journal articles from the early days of genetics. Sometimes I daydreamed about discovering the secret of the gene, but not once did I have the faintest trace of a respectable idea. It was thus difficult to avoid the disquieting thought that I was not accomplishing anything. Knowing that I had not come to Naples for work did not make me feel better.

I retained a slight hope that I might profit from the meeting on the structures of biological macromolecules. Though I knew nothing about the X-ray diffraction techniques that dominated structural analysis, I was optimistic that the spoken arguments would be more comprehensible than the journal articles, which passed over my head. I was specially interested to hear the talk on nucleic acids to be given by Randall. At that time almost nothing was published about the possible three-dimensional configurations of a nucleic-acid molecule. Conceivably this fact affected my casual pursuit of chemistry. For why should I get excited learning boring chemical facts as long as the chemists never provided anything incisive about the nucleic acids?

The odds, however, were against any real revelation then. Much of the talk about the three-dimensional structure of proteins and nucleic acids was hot air. Though this work had been going on for over fifteen years, most if not all of the facts were soft. Ideas put forward with conviction were likely to be the products of wild crystallographers who delighted in being in a field where their ideas could not be easily disproved. Thus, although virtually all biochemists, including Herman, were unable to understand the arguments of the X-ray people, there was little uneasiness. It made no sense to learn complicated mathematical methods in order to follow baloney. As a result, none of my teachers had ever considered the possibility that I might do postdoctoral research with an X-ray crystallographer.

Maurice, however, did not disappoint me. The fact that he was a substitute for Randall made no difference: I had not known about either. His talk was far from vacuous and stood out sharply from the rest, several of which bore no connection to the purpose of the meeting. Fortunately these were in Italian, and so the obvious boredom of the foreign guests did not need to be construed as impoliteness. Several other speakers were continental biologists, at that time guests at the Zoological Station, who only briefly alluded to macromolecular structure. In contrast, Maurice's X-ray diffraction picture of DNA was to the point. It was flicked on the screen near the end of his talk. Maurice's dry English form did not permit enthusiasm as he stated that the picture showed much more detail than previous pictures and could, in fact, be considered as arising from a crystalline substance. And when the structure of DNA was known, we might be in a better position to understand how genes work.

Suddenly I was excited about chemistry. Before Maurice's talk I had worried about the possibility that the gene might be fantastically irregular. Now, however, I knew that genes could crystallize; hence they must have a regular structure that could be solved in a straightforward fashion. Immediately I began to wonder whether it would be possible for me to join Wilkins in working on DNA. After the lecture I tried to seek him out. Perhaps he already knew more than his talk had indicated—often if a scientist is not absolutely sure he is correct, he is hesitant to speak in public. But there was no opportunity to talk to him; Maurice had vanished.

Not until the next day, when all the participants took an excursion to the Greek temples at Paestum, did I get an opportunity to introduce myself. While waiting for the bus I started a conversation and explained how interested I was in DNA. But before I could pump Maurice we had to board, and I joined my sister, Elizabeth, who had just come in from the States. At the temples we all scattered, and before I could corner Maurice again I realized that I might have had a tremendous stroke of good luck. Maurice had noticed that my sister was very pretty, and soon they were eating lunch together. I was immensely pleased. For years I had sullenly watched Elizabeth being pursued by a series of dull nitwits. Suddenly the possibility opened up that her way of life could be changed. No longer did I have to face the certainty that she would end up with a mental defective. Furthermore, if Maurice really liked my sister, it was inevitable that I would become closely associated with his X-ray work on DNA. The fact that Maurice excused himself to go and sit alone did not upset me. He obviously had good manners and assumed that I wished to converse with Elizabeth.

As soon as we reached Naples, however, my daydreams of glory by association ended. Maurice moved off to his hotel with only a casual nod. Neither the beauty of my sister nor my intense interest in the DNA structure had snared him. Our futures did not seem to be in London. Thus I set off to Copenhagen and the prospect of more biochemistry to avoid.

5

I PROCEEDED to forget Maurice, but not this DNA photograph. A potential key to the secret of life was impossible to push out of my mind. The fact that I was unable to interpret it did not bother me. It was certainly better to imagine myself becoming famous than maturing into a stifled academic who had never risked a thought. I was also encouraged by the very exciting rumor that Linus Pauling had partly solved the structure of proteins. The news hit me in Geneva, where I had stopped for several days to talk with the Swiss phage worker Jean Weigle, who was just back from a winter of work at Cal Tech. Before leaving, Jean had gone to the lecture where Linus had made the announcement.

Pauling's talk was made with his usual dramatic flair. The words came out as if he had been in show business all his life. A curtain kept his model hidden until near the end of his lecture, when he proudly unveiled his latest creation. Then, with his eyes twinkling, Linus explained the specific characteristics that made his model—the α-helix—uniquely beautiful. This show, like all of his dazzling performances, delighted the younger students in attendance. There was no one like Linus in all the world. The combination of his prodigious mind and his infectious grin was unbeatable. Several fellow professors, however, watched this performance with mixed feelings. Seeing Linus jumping up and down on the demonstration table and moving his arms like a magician about to pull a rabbit out of his shoe made them feel inadequate. If only he had shown a little humility, it would have been so much easier to take! Even if he were to say nonsense, his mesmerized students would never know because of his unquenchable self-confidence. A number of his colleagues quietly waited for the day when he would fall flat on his face by botching something important.

But Jean could not tell me whether Linus' α-helix

was right. He was not an X-ray crystallographer and could not judge the model professionally. Several of his younger friends, however, trained in structural chemistry, thought the α-helix looked very pretty. The best guess of Jean's acquaintances, therefore, was that Linus was right. If so, he had again accomplished a feat of extraordinary significance. He would be the first person to propose something solidly correct about the structure of a biologically important macromolecule. Conceivably, in doing so, he might have come up with a sensational new method which could be extended to the nucleic acids. Jean, however, did not remember any special tricks. The most he could tell me was that a description of the α-helix would soon be published.

By the time I was back in Copenhagen, the journal containing Linus' article had arrived from the States. I quickly read it and immediately reread it. Most of the language was above me, and so I could only get a general impression of his argument. I had no way of judging whether it made sense. The only thing I was sure of was that it was written with style. A few days later the next issue of the journal arrived, this time containing seven more Pauling articles. Again the language was dazzling and full of rhetorical tricks. One article started with the phrase, "Collagen is a very interesting protein." It inspired me to compose opening lines of the paper I would write about DNA, if I solved its structure. A sentence like "Genes are interesting to geneticists" would distinguish my way of thought from Pauling's.

So I began worrying about where I could learn how to solve X-ray diffraction pictures. Cal Tech was not the place—Linus was too great a man to waste his time teaching a mathematically deficient biologist. Neither did I wish to be further put off by Wilkins. This left Cambridge, England, where I knew that someone named Max Perutz was interested in the structure of the large biological molecules, in particular, the protein hemoglobin. I thus wrote to Luria about my newly found passion, asking whether he knew how to arrange my acceptance into the Cambridge lab. Unexpectedly, this was no problem at all. Soon after receiving my letter, Luria went to a small meeting at Ann Arbor, where he met Perutz' coworker, John Kendrew, then on an extended trip to the States.

Most fortunately, Kendrew made a favorable impression on Luria; like Kalckar, he was civilized and in addition supported the Labor Party. Furthermore, the Cambridge lab was understaffed and Kendrew was looking for someone to join him in his study of the protein myoglobin. Luria assured him that I would fit the bill and immediately wrote me the good news.

It was then early August, just a month before my original fellowship would expire. This meant that I could not long delay writing to Washington about my change of plans. I decided to wait until I was admitted officially into the Cambridge lab. There was always the possibility that something would go wrong. It seemed prudent to put off the awkward letter until I could talk personally with Perutz. Then I could state in much greater detail what I might hope to accomplish in England. I did not, however, leave at once. Again I was back in the lab, and the experiments I was doing were fun, in a second-class fashion. Even more important, I did not want to be away during the forthcoming International Poliomyelitis Conference, which was to bring several phage workers to Copenhagen. Max Delbrück was in the expected group, and since he was a professor at Cal Tech he might have further news about Pauling's latest trick.

Delbrück, however, did not enlighten me further. The α-helix, even if correct, had not provided any biological insights; he seemed bored speaking about it. Even my information that a pretty X-ray photograph of DNA existed elicited no real response. But I had no opportunity to be depressed by Delbrück's characteristic bluntness, for the poliomyelitis congress was an unparalleled success. From the moment the several hundred delegates arrived, a profusion of free champagne, partly provided by American dollars, was available to loosen international barriers. Each night for a week there were receptions, dinners, and midnight trips to waterfront bars. It was my first experience with the high life, associated in my mind with decaying European aristocracy. An important truth was slowly entering my head: a scientist's life might be interesting socially as well as intellectually. I went off to England in excellent spirits.

6

MAX PERUTZ was in his office when I showed up just after lunch. John Kendrew was still in the States, but my arrival was not unexpected. A brief letter from John said that an American biologist might work with him during the following year. I explained that I was ignorant of how X rays diffract, but Max immediately put me at ease. I was assured that no high-powered mathematics would be required: both he and John had studied chemistry as undergraduates. All I need do was read a crystallographic text; this would enable me to understand enough theory to begin to take X-ray photographs. As an example, Max told me about his simple idea for testing Pauling's α-helix. Only a day had been required to get the crucial photograph confirming Pauling's prediction. I did not follow Max at all. I was even ignorant of Bragg's Law, the most basic of all crystallographic ideas.

We then went for a walk to look over possible digs for the coming year. When Max realized that I had come directly to the lab from the station and had not yet seen any of the colleges, he altered our course to take me through King's, along the backs, and through to the Great Court of Trinity. I had never seen such beautiful buildings in all my life, and any hesitation I might have had about leaving my safe life as a biologist vanished. Thus I was only nominally depressed when I peered inside several damp houses known to contain student rooms. I knew from the novels of Dickens that I would not suffer a fate the English denied themselves. In fact, I thought myself very lucky when I found a room in a two-story house on Jesus Green, a superb location less than ten minutes' walk from the lab.

The following morning I went back to the Cavendish, since Max wanted me to meet Sir Lawrence Bragg. When Max telephoned upstairs that I was here, Sir Lawrence came down from his office, let me say a few words, and then retired for a private conversation with Max. A few minutes later they emerged to allow Bragg to give me his formal permission to work under his direction. The performance was uncompromisingly British, and I quietly concluded that the white-mustached figure of Bragg now spent most of its days sitting in London clubs like the Athenaeum.

The thought never occurred to me then that later on I would have contact with this apparent curiosity out of the past. Despite his indisputable reputation, Bragg had worked out his Law just before World War I, so I assumed he must be in effective retirement and would never care about genes. I politely thanked Sir Lawrence for accepting me and told Max I would be back in three weeks for the start of the Michaelmas term. I then returned to Copenhagen to collect my few clothes and to tell Herman about my good luck in being able to become a crystallographer.

Herman was splendidly cooperative. A letter was dispatched telling the Fellowship Office in Washington that he enthusiastically endorsed my change in plans. At the same time I wrote a letter to Washington, breaking the news that my current experiments on the biochemistry of virus reproduction were at best interesting in a nonprofound way. I wanted to give up conventional biochemistry, which I believed incapable of telling us how genes work. Instead I told them that I now knew that X-ray crystallography was the key to genetics. I requested the approval of my plans to transfer to Cambridge so that I might work at Perutz' lab and learn how to do crystallographic research.

I saw no point in remaining in Copenhagen until permission came. It would have been absurd to stay there wasting my time. The week before, Maaløe had departed for a year at Cal Tech, and my interest in Herman's type of biochemistry remained zero. Leaving Copenhagen was of course illegal in the formal sense. On the other hand, my request could not be refused. Everyone knew of Herman's unsettled state, and the Washington office must have been wondering how long I would care to remain in Copenhagen. Writing directly about Herman's absence from his lab would have been not only ungentlemanly but unnecessary.

Naturally I was not at all prepared to receive a letter refusing permission. Ten days after my return to Cambridge, Herman forwarded the depressing news, which had been sent to my Copenhagen address. The Fellowship Board would not approve my transfer to a lab from which I was totally unprepared to profit. I was told to reconsider my plans, since I was unqualified to do crystallographic work. The Fellowship Board would, however, look favorably on a proposal that I transfer to the cell-physiology laboratory of Caspersson in Stockholm.

The source of the trouble was all too apparent. The head of the Fellowship Board no longer was Hans Clarke, a kindly biochemist friend of Herman's, then about to retire from Columbia. My letter had gone instead to a new chairman, who took a more active interest in directing young people. He was put out that I had overstepped myself in denying that I would profit from biochemistry. I wrote to Luria to save me. He and the new man were casual acquaintances, and so when my decision was set in proper perspective, he might reverse his decision.

At first there were hints that Luria's interjection might cause a change back to reason. I was cheered up when a letter arrived from Luria that the situation might be smoothed over if we appeared to eat crow. I was to write Washington that a major inducement in my wanting to be in Cambridge was the presence of Roy Markham, an English biochemist who worked with plant viruses. Markham took the news quite casually when I walked into his office and told him that he might acquire a model student who would never bother him by cluttering up his lab with experimental apparatus. He regarded the scheme as a perfect example of the inability of Americans to know how to behave. Nonetheless, he promised to go along with this nonsense.

Armed with the assurance that Markham would not squeal, I humbly wrote a long letter to Washington, outlining how I might profit from being in the joint presence of Perutz and Markham. At the end of the letter I thought it honest to break the news officially that I was in Cambridge and would remain there until a decision was made. The new man in Washington, however, did not play ball. The clue came when the return letter was addressed to Herman's lab. The Fellowship Board was considering my case. I would be informed when a decision had been made. Thus it did not seem prudent to cash my checks, which were still sent to Copenhagen at the beginning of each month.

Fortunately, the possibility of my not being paid in the forthcoming year for working on DNA was only annoying and not fatal. The $3000 fellowship stipend that I had received for being in Copenhagen was three times that required to live like a well-off Danish student. Even if I had to cover my sister's recent purchase of two fashionable Paris suits, I would have $1000 left, enough for a year's stay in Cambridge. My landlady was also helpful. She threw me out after less than a month's residence. My main crime was not removing my shoes when I entered the house after 9:00 P.M., the hour at which her husband went to sleep. Also I occasionally forgot the injunction not to flush the toilet at similar hours and, even worse, I went out after 10:00 P.M. Nothing in Cambridge was then open, and my motives were suspect. John and Elizabeth Kendrew rescued me with the offer, at almost no rent, of a tiny room in their house on Tennis Court Road. It was unbelievably damp and heated only by an aged electric heater. Nonetheless, I eagerly accepted the offer. Though it looked like an open invitation to tuberculosis, living with friends was infinitely preferable to any other digs I might find at this late moment. So without any reluctance I decided to stay at Tennis Court Road until my financial picture improved.

7

FROM my first day in the lab I knew I would not leave Cambridge for a long time. Departing would be idiocy, for I had immediately discovered the fun of talking to Francis Crick. Finding someone in Max's lab who knew that DNA was more important than proteins was real luck. Moreover, it was a great relief for me not to spend full time learning X-ray analysis of proteins. Our lunch conversations quickly centered on how genes were put together. Within a few days after my arrival, we knew what to do: imitate Linus Pauling and beat him at his own game.

Pauling's success with the polypeptide chain had naturally suggested to Francis that the same tricks might also work for DNA. But as long as no one nearby thought DNA was at the heart of everything, the potential personal difficulties with the King's lab kept him from moving into action with DNA. Moreover, even though hemoglobin was not the center of the universe, Francis' previous two years at the Cavendish certainly had not been dull. More than enough protein problems kept popping up that required someone with a bent toward theory. But now, with me around the lab always wanting to talk about genes, Francis no longer kept his thoughts about DNA in a back recess of his brain. Even so, he had no intention of abandoning his interest in the other laboratory problems. No one should mind if, by spending only a few hours a week thinking about DNA, he helped me solve a smashingly important problem.

As a consequence, John Kendrew soon realized that I was unlikely to help him solve the myoglobin structure. Since he was unable to grow large crystals of horse myoglobin, he initially hoped I might have a greener thumb. No effort, however, was required to see that my laboratory manipulations were less skillful than those of a Swiss chemist. About a fortnight after my arrival in Cambridge, we went out to the local slaughterhouse to get a horse heart for a new myoglobin preparation. If we were lucky, the damage to the myoglobin molecules which prevented crystallization would be averted by immediately freezing the ex-racehorse's heart. But my subsequent attempts at crystallization were no more successful than John's. In a sense I was almost relieved. If they had succeeded, John might have put me onto taking X-ray photographs.

No obstacle thus prevented me from talking at least several hours each day to Francis. Thinking all the time was too much even for Francis, and often when he was stumped by his equations he used to pump my reservoir of phage lore. At other moments Francis would endeavor to fill my brain with cyrstallographic facts, ordinarily available only through the painful reading of professional journals. Particularly important were the exact arguments needed to understand how Linus Pauling had discovered the α-helix.

I soon was taught that Pauling's accomplishment was a product of common sense, not the result of complicated mathematical reasoning. Equations occasionally crept into his argument, but in most cases words would have sufficed. The key to Linus' success was his reliance on the simple laws of structural chemistry. The α-helix had not been found by only staring at X-ray pictures; the essential trick, instead, was to ask which atoms like to sit next to each other. In place of pencil and paper, the main working tools were a set of molecular models superficially resembling the toys of preschool children.

We could thus see no reason why we should not solve DNA in the same way. All we had to do was to construct a set of molecular models and begin to play—with luck, the structure would be a helix. Any other type of configuration would be much more complicated. Worrying about complications before ruling out the possiblity that the answer was simple would have been damned foolishness. Pauling never got anywhere by seeking out messes.

From our first conversations we assumed that the DNA molecule contained a very large number of nucleotides linearly linked together in a regular way. Again our reasoning was partially based upon simplicity. Although organic chemists in Alexander Todd's nearby lab thought this the basic arrangement, they were still a long way from chemically establishing that all the internucleotide bonds were identical. If this was not the case, however, we could not see how the DNA molecules packed together to form the crystalline aggregates studied by Maurice Wilkins and Rosalind Franklin. Thus, unless we found all future progress blocked, the best course was to regard the sugar-phosphate backbone as extremely regular and to search for a helical three-dimensional configuration in which all the backbone groups had identical chemical environments.

Immediately we could see that the solution to DNA might be more tricky than that of the α-helix. In the α-helix, a single polypeptide (a collection of amino acids) chain folds up into a helical arrangement held together by hydrogen bonds between groups on the same chain. Maurice had told Francis, however, that the diameter of the DNA molecule was thicker than would be the case of only one polynucleotide (a collection of nucleotides) chain were present. This made him think that the DNA molecule was a compound helix composed of several polynucleotide chains twisted about each other. If true, then before serious model building began, a decision would have to be made whether the chains would be held together by hydrogen bonds or by salt linkages involving the negatively charged phosphate groups.

A further complication arose from the fact that four types of nucleotides were found in DNA. In this sense, DNA was not a regular molecule but a highly irregular one. The four nucleotides were not, however, completely different, for each contained the same sugar and phosphate components. Their uniqueness lay in their nitrogenous bases, which were either a purine (adenine and guanine) or a pyrimidine (cytosine and thymine). But since the linkages between the nucleotides involved only the phosphate and sugar groups, our assumption that the same type of chemical bond linked all the nucleotides together was not affected. So in building models we would postulate that the sugar-phosphate backbone was very regular, and the order of bases of necessity very irregular. If the base sequences were always the same, all DNA molecules would be identical and there would not exist the variability that must distinguish one gene from another.

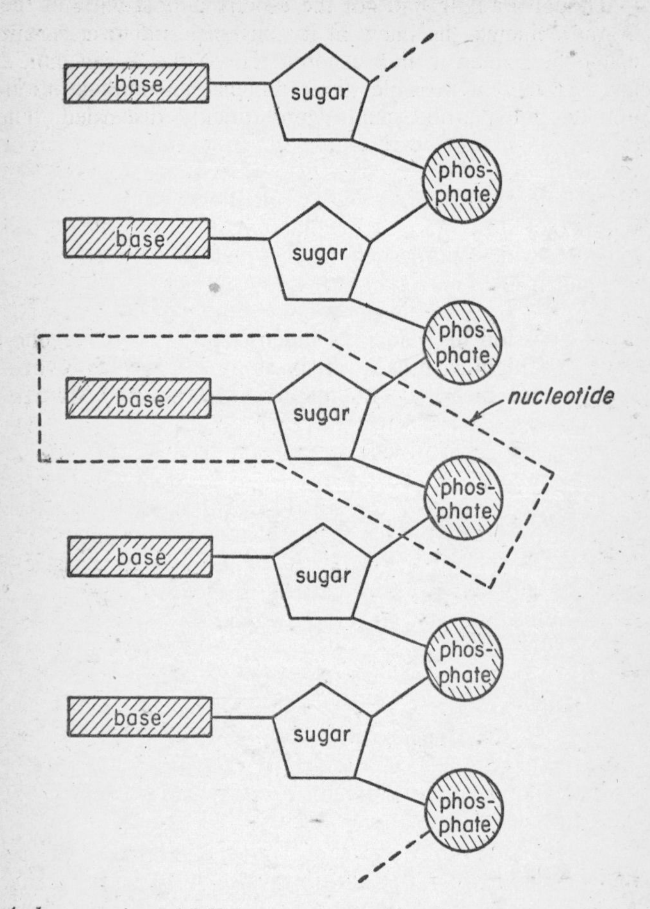

A short section of DNA as envisioned by Alexander Todd's research group in 1951. They thought that all the internucleotide links were phosphodiester bonds joining sugar carbon atom #5 to sugar carbon atom #3 of the adjacent nucleotide. As organic chemists they were concerned with how the atoms were linked together, leaving to crystallographers the problem of the 3-D arrangement of the atoms.

Though Pauling had got the α-helix almost without the X-ray evidence, he knew of its existence and to a certain extent had taken it into account. Given the X-ray data, a large variety of possible three-dimensional configurations for the polypeptide chain were quickly discarded. The exact X-ray data should help us go ahead much faster with the more subtly constructed DNA molecule. Mere inspection of the DNA X-ray picture should prevent a number of false starts. Fortunately, there already existed one half-good photograph in the published literature. It was taken five years previously by the English crystallographer W. T. Astbury, and could be used to start us off. Yet possession of Maurice's much better crystalline photograph's might save us from six months' to a year's work. The painful fact that the pictures belonged to Maurice could not be avoided.

There was nothing else to do but talk to him. To our surprise, Francis had no problem in persuading Maurice to come up to Cambridge for a weekend. And there was no need to force Maurice to the conclusion that the structure was a helix. Not only was it the obvious guess, but Maurice already had been talking in terms of helices at a summer meeting in Cambridge. About six weeks before I first arrived there, he had shown X-ray diffraction pictures of DNA which revealed a marked absence of reflections on the meridian. This was a feature that his colleague, the theoretician Alex Stokes, had told him was compatible with a helix. Given this conclusion, Maurice suspected that three polynucleotide chains were used to construct the helix.

He did not, however, share our belief that Pauling's model-building game would quickly solve the structure, at least not until further X-ray results were obtained. Most of our conversation, instead, centered on Rosy Franklin. More trouble than ever was coming from her. She was now insisting that not even Maurice himself should take any more X-ray photographs of DNA. In trying to come to terms with Rosy, Maurice made a very bad bargain. He had handed over to her all the good crystalline DNA used in his original work and had agreed to confine his studies to other DNA, which he afterward found did not crystallize.

PURINES

PYRIMIDINES

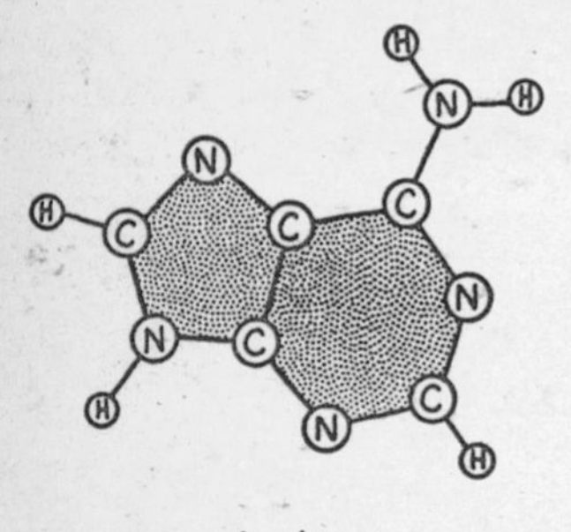

adenine

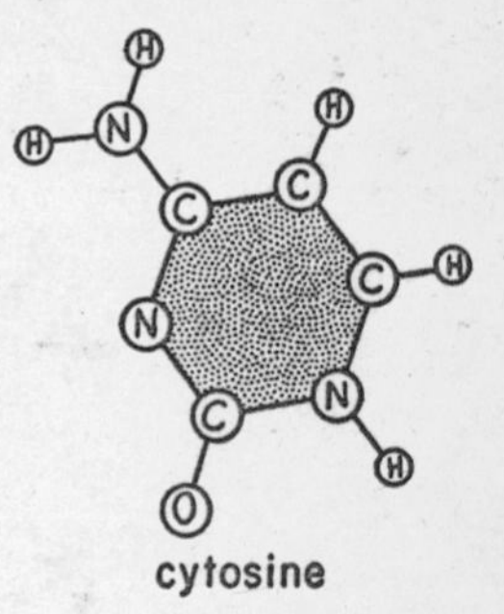

cytosine

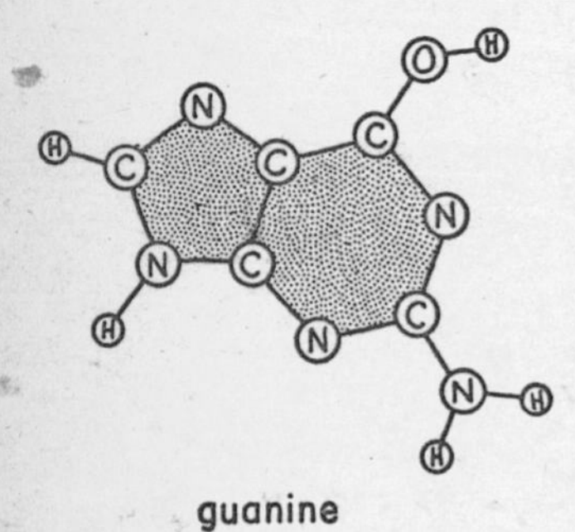

guanine

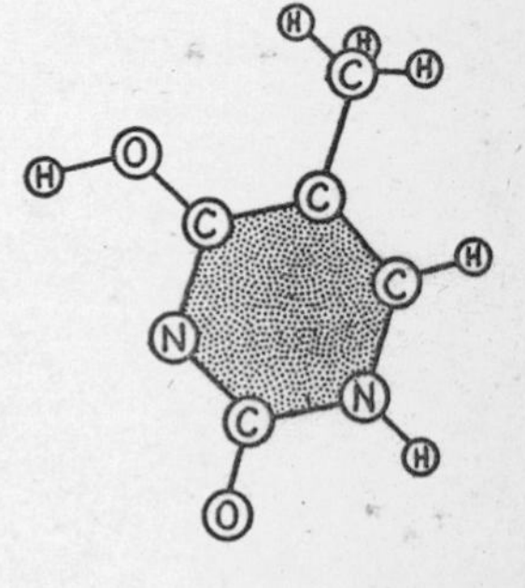

thymine

The chemical structures of the four DNA bases as they were often drawn about 1951. Because the electrons in the five- and six-membered rings are not localized, each base has a planar shape with a thickness of 3.4 Å.

The point had been reached where Rosy would not even tell Maurice her latest results. The soonest Maurice was likely to learn where things stood was three weeks hence, the middle of November. At that time Rosy was scheduled to give a seminar on her past six months' work. Naturally I was delighted when Maurice said I would be welcome at Rosy's talk. For the first time I had a real incentive to learn some crystallography: I did not want Rosy to speak over my head.

8

MOST unexpectedly, Francis' interest in DNA temporarily fell to almost zero less than a week later. The cause was his decision to accuse a colleague of ignoring his ideas. The accusation was leveled at none other than his Professor. It happened less than a month after my arrival, on a Saturday morning. The previous day Max Perutz had given Francis a new manuscript by Sir Lawrence and himself, dealing with the shape of the hemoglobin molecule. As he rapidly read its contents Francis became furious, for he noticed that part of the argument depended upon a theoretical idea he had propounded some nine months earlier. What was worse, Francis remembered having enthusiastically proclaimed it to everyone in the lab. Yet his contribution had not been acknowledged. Almost at once, after dashing in to tell Max and John Kendrew about the outrage, he hurried to Bragg's office for an explanation, if not an apology. But by then Bragg was at home, and Francis had to wait until the following morning. Unfortunately, this delay did not make the confrontation any more successful.

Sir Lawrence flatly denied prior knowledge of Francis' efforts and was thoroughly insulted by the implication that he had underhandedly used another scientist's ideas. On the other hand, Francis found it impossible to believe that Bragg could have been so dense as to have missed his oft-repeated idea, and he as much as told Bragg this. Further conversation became impossible, and in less than ten minutes Francis was out of the Professor's office.

For Bragg this meeting seemed the final straw in his relations with Crick. Several weeks earlier Bragg had come into the lab greatly excited about an idea that came to him the previous evening, one that he and Perutz subsequently incorporated in their paper. While he was explaining it to Perutz and Kendrew, Crick happened to join the group. To his considerable annoyance, Francis did not accept the idea immediately but instead stated that he would go away and check whether Bragg was right or wrong. At this stage Bragg had blown his top and, with his blood pressure all too high, returned home presumably to tell his wife about the latest antics of their problem child.

This most recent tussle was a disaster for Francis, and he showed his uneasiness when he came down to the lab. Bragg, in dismissing him from his room, had angrily told him that he would consider seriously whether he could continue to give Francis a place in the laboratory after his Ph.D. course was ended. Francis was obviously worried that he might soon have to find a new position. Our subsequent lunch at the Eagle, the pub at which he usually ate, was restrained and unpunctuated by the usual laughter.

His concern was not without reason. Although he knew he was bright and could produce novel ideas, he could claim no clear-cut intellectual achievements and he was still without his Ph.D. He came from a solid middle-class family and was sent to school at Mill Hill. Then he read physics at University College, London, and had commenced work on an advanced degree when the war broke out. Like almost all other English scientists, he joined the war effort and became part of the Admiralty's scientific establishment. There he worked with great vigor, and, although many resented his nonstop conversation, there was a war to win and he was quite helpful in producing ingenious magnetic mines. When the war was over, however, some of his colleagues saw no sound reason to have him about forever, and for a period he was given to believe that he had no future in the scientific civil service.

Moreover, he had lost all desire to stay in physics and decided instead to try biology. With the help of the physiologist A. V. Hill, he obtained a small grant to come up to Cambridge in the fall of 1947. At first he did true biology at the Strangeways Laboratory, but this was obviously trivial and two years later he moved over to the Cavendish, where he joined Perutz and Kendrew. Here he again became excited about science and decided that perhaps he should finally work for a Ph.D. He thus enrolled as a research student (of Caius College) with Max as his supervisor. In a sense, this pursuit of the Ph.D. was a bore to a mind that worked too fast to be satisfied with the tedium involved in thesis research. On the other hand, his decision had yielded an unforeseen dividend: in this moment of crisis, he could hardly be dismissed before he got his degree.

Max and John quickly came to Francis' rescue and interceded with the Professor. John confirmed that Francis had previously written an account of the argument in question, and Bragg acknowledged that the same idea had occurred independently to both. Bragg by that time had calmed down, and any question of Crick's going was quietly shelved. Keeping him on was not easy on Bragg. One day, in a moment of despair, he revealed that Crick made his ears buzz. Moreover, he remained unconvinced that Crick was needed. Already for thirty-five years he had not stopped talking and almost nothing of fundamental value had emerged.

9

A NEW opportunity to theorize soon brought Francis back to normal form. Several days after the fiasco with Bragg, the crystallographer V. Vand sent Max a letter containing a theory for the diffraction of X rays by helical molecules. Helices were then at the center of the lab's interest, largely because of Pauling's α-helix. Yet there was still lacking a general theory to test new models as well as to confirm the finer details of the α-helix. This is what Vand hoped his theory would do.

Francis quickly found a serious flaw in Vand's efforts, became excited about finding the right theory, and bounded upstairs to talk with Bill Cochran, a small, quiet Scot, then a lecturer in crystallography at the Cavendish. Bill was the cleverest of the younger Cambridge X-ray people, and even though he was not involved in work on the large biological macromolecules, he always provided the most astute sounding board for Francis' frequent ventures into theory. When Bill told Francis that an idea was unsound or would lead nowhere, Francis could be sure that professional jealousy was not involved. This time, however, Bill did not voice skepticism, since independently he had found faults in Vand's paper and had begun to wonder what the right answer was. For months both Max and Bragg had been after him to work out the helical theory, but he had not moved into action. Now, with the additional pressure from Francis, he too began seriously to ponder how the equations should be set up.

The remainder of the morning Francis was silent and absorbed in mathematical equations. At lunch at the Eagle a bad headache came on, and he went home instead of returning to the lab. But sitting in front of the gas fire doing nothing bored him, and again he took up his equations. To his delight, he soon saw that he had the answer. Nonetheless, he stopped his work, for he and his wife, Odile, were invited to a wine tasting at Matthews', one of Cambridge's better wine merchants. For several days his morale had been buoyed by the request to sample the wines. It meant acceptance by a more fashionable and amusing part of Cambridge and allowed him to dismiss the fact that he was not appreciated by a variety of dull and pompous dons.

He and Odile were then living at the "Green Door," a tiny, inexpensive flat on top of a several-hundred-year-old house just across Bridge Street from St. John's College. There were only two rooms of any size, a livingroom and a bedroom. All the others, including the kitchen, in which the bathtub was the largest and most conspicuous object, were almost nonexistent. But despite the cramp, its great charm, magnified by Odile's decorative sense, gave it a cheerful, if not playful, spirit. Here I first sensed the vitality of English intellectual life, so completely absent during my initial days in my Victorian room several hundred yards away on Jesus Green.

They had then been married for three years. Francis' first marriage did not last long, and a son, Michael, was looked after by Francis' mother and aunt. He had lived alone for several years until Odile, some five years his junior, came to Cambridge and hastened his revolt against the stodginess of the middle classes, which delight in unwicked amusements like sailing and tennis, habits particularly unsuited to the conversational life. Neither was politics or religion of any concern. The latter was clearly an error of past generations, which Francis saw no reason to perpetuate. But I am less certain about their complete lack of enthusiasm for political issues. Perhaps it was the war, whose grimness they now wished to forget. In any case, *The Times* was not present at breakfast, and more attention was given to *Vogue,* the only magazine to which they subscribed and about which Francis could converse at length.

By then I was often going to the Green Door for dinner. Francis was always eager to continue our conversations, while I joyously seized every opportunity to escape from the miserable English food that periodically led me to worry about whether I might have an ulcer. Odile's French mother had imparted to her a thorough contempt for the unimaginative way in which most Englishmen eat and house themselves. Francis thus never had reason to envy those college fellows whose High Table food was undeniably better than their wives' drab mixtures of tasteless meat, boiled potatoes, colorless greens, and typical trifles. Instead, dinner was often gay, especially after the wine turned the conversation to the currently talked-about Cambridge popsies.

There was no restraint in Francis' enthusiasms about young women—that is, as long as they showed some vitality and were distinctive in any way that permitted gossip and amusement. When young, he saw little of women and was only now discovering the sparkle they added to life. Odile did not mind this predilection, seeing that it went along with, and probably helped, his emancipation from the dullness of his Northampton upbringing. They would talk at length about the somewhat artsy-craftsy world in which Odile moved and into which they were frequently invited. No choice event was kept out of our conversations, and he would show equal gusto in telling of his occasional mistakes. One occurred when there was a costume party and he went looking like G. B. Shaw in a full red beard. As soon as he entered he realized that it was a ghastly error, since not one of the young women enjoyed being tickled by the wet, scraggly hairs when he came within kissing distance.

But there were no young women at the wine tasting. To his and Odile's dismay, their companions were college dons contentedly talking about the burdensome administrative problems with which they were so sadly afflicted. They went home early and Francis, unexpectedly sober, thought more about his answer.

The next morning he arrived in the lab and told Max and John about his success. A few minutes later, Bill Cochran walked into his office, and Francis started to repeat the story. But before he could let loose his argument, Bill told him that he also thought he had succeeded. Hurriedly they went through their respective mathematics and discovered that Bill had used an elegant derivation compared to Francis' more laborious approach. Gleefully, however, they found that they had arrived at the same final answer. They then checked the α-helix by visual inspection with Max's X-ray diagrams. The agreement was so good that both Linus' model and their theory had to be correct.

Within a few days a polished manuscript was ready and jubilantly dispatched to *Nature*. At the same time, a copy was sent to Pauling to appreciate. This event, his first unquestionable success, was a signal triumph for Francis. For once the absence of women had gone along with luck.

10

BY mid-November, when Rosy's talk on DNA rolled about, I had learned enough crystallographic argument to follow much of her lecture. Most important, I knew what to focus attention upon. Six weeks of listening to Francis had made me realize that the crux of the matter was whether Rosy's new X-ray pictures would lend any support for a helical DNA structure. The really relevant experimental details were those which might provide clues in constructing molecular models. It took, however, only a few minutes of listening to Rosy to realize that her determined mind had set upon a different course of action.

She spoke to an audience of about fifteen in a quick, nervous style that suited the unornamented old lecture hall in which we were seated. There was not a trace of warmth or frivolity in her words. And yet I could not regard her as totally uninteresting. Momentarily I wondered how she would look if she took off her glasses and did something novel with her hair. Then, however, my main concern was her description of the crystalline X-ray diffraction pattern.

The years of careful, unemotional crystallographic training had left their mark. She had not had the advantage of a rigid Cambridge education only to be so foolish as to misuse it. It was downright obvious to her that the only way to establish the DNA structure was by pure crystallographic approaches. As model building did not appeal to her, at no time did she mention Pauling's triumph over the α-helix. The idea of using tinker-toy-like models to solve biological structures was clearly a last resort. Of course Rosy knew of Linus' success but saw no obvious reason to ape his mannerisms. The measure of his past triumphs was sufficient reason in itself to act differently; only a genius of his stature could play like a ten-year-old boy and still get the right answer.

Rosy regarded her talk as a preliminary report which,

by itself, would not test anything fundamental about DNA. Hard facts would come only when further data had been collected which could allow the crystallographic analyses to be carried to a more refined stage. Her lack of immediate optimism was shared by the small group of lab people who came to the talk. No one else brought up the desirability of using molecular models to help solve the structure. Maurice himself only asked several questions of a technical nature. The discussion then quickly stopped with the expressions on the listeners' faces indicating either that they had nothing to add or that, if they did wish to say something, it would be bad form since they had said it before. Maybe their reluctance to utter anything romantically optimistic, or even to mention models, was due to fear of a sharp retort from Rosy. Certainly a bad way to go out into the foulness of a heavy, foggy November night was to be told by a woman to refrain from venturing an opinion about a subject for which you were not trained. It was a sure way of bringing back unpleasant memories of lower school.

Following some brief and, as I was later to observe, characteristically tense small talk with Rosy, Maurice and I walked down the Strand and across to Choy's Restaurant in Soho. Maurice's mood was surprisingly jovial. Slowly and precisely he detailed how, in spite of much elaborate crystallographic analysis, little real progress had been made by Rosy since the day she arrived at King's. Though her X-ray photographs were somewhat sharper than his, she was unable to say anything more positive than he had already. True, she had done some more detailed measurements of the water content of her DNA samples, but even here Maurice had doubts about whether she was really measuring what she claimed.

To my surprise, Maurice seemed buoyed up by my presence. The aloofness that existed when we first met in Naples had vanished. The fact that I, a phage person, found what he was doing important was reassuring. It really was no help to receive encouragement from a fellow physicist. Even when he met those who thought his decision to go into biology made sense, he couldn't trust their judgment. After all, they didn't know any biology, and so it was best to take their remarks as politeness, even condescension, toward someone opposed to the competitive pace of postwar physics.

To be sure, he got active and very necessary help from some biochemists. If not, he could never have come into the game. Several of them had been absolutely vital in generously providing him with samples of highly purified DNA. It was bad enough learning crystallography without having to acquire the witchcraft-like techniques of the biochemist. On the other hand, the majority weren't like the high-powered types he had worked with on the bomb project. Sometimes they seemed even ignorant of the way DNA was important.

But even so they knew more than the majority of biologists. In England, if not everywhere, most botanists and zoologists were a muddled lot. Not even the possession of University Chairs gave many the assurance to do clean science; some actually wasted their efforts on useless polemics about the origin of life or how we know that a scientific fact is really correct. What was worse, it was possible to get a university degree in biology without learning any genetics. That was not to say that the geneticists themselves provided any intellectual help. You would have thought that with all their talk about genes they should worry about what they were. Yet almost none of them seemed to take seriously the evidence that genes were made of DNA. This fact was unnecessarily chemical. All that most of them wanted out of life was to set their students onto uninterpretable details of chromosome behavior or to give elegantly phrased, fuzzy-minded speculations over the wireless on topics like the role of the geneticist in this transitional age of changing values.

So the knowledge that the phage group took DNA seriously made Maurice hope that times would change and he would not have painfully to explain, each time he gave a seminar, why his lab was making so much fuss and bother about DNA. By the time our dinner was finished, he was clearly in a mood to push ahead. Yet all too suddenly Rosy popped back into the conversation, and the possibility of really mobilizing his lab's efforts slowly receded as we paid the bill and went out into the night.

11

THE following morning I joined Francis at Paddington Station. From there we were to go up to Oxford to spend the weekend. Francis wanted to talk to Dorothy Hodgkin, the best of the English crystallographers, while I welcomed the opportunity to see Oxford for the first time. At the train gate Francis was in top form. The visit would give him the opportunity to tell Dorothy about his success with Bill Cochran in working out the helical diffraction theory. The theory was much too elegant not to be told in person—individuals like Dorothy who were clever enough to understand its power immediately were much too rare.

As soon as we were in the train carriage, Francis began asking questions about Rosy's talk. My answers were frequently vague, and Francis was visibly annoyed by my habit of always trusting to memory and never writing anything on paper. If a subject interested me, I could usually recollect what I needed. This time, however, we were in trouble, because I did not know enough of the crystallographic jargon. Particularly unfortunate was my failure to be able to report exactly the water content of the DNA samples upon which Rosy had done her measurements. The possibility existed that I might be misleading Francis by an order-of-magnitude difference.

The wrong person had been sent to hear Rosy. If Francis had gone along, no such ambiguity would have existed. It was the penalty for being oversensitive to the situation. For, admittedly, the sight of Francis mulling over the consequences of Rosy's information when it was hardly out of her mouth would have upset Maurice. In one sense it would be grossly unfair for them to learn the facts at the same time. Certainly Maurice should have the first chance to come to grips with the problem. On the other hand, there seemed no indication that he thought the answer would come from playing with molecular models. Our conversation on the previous night had hardly alluded to that approach. Of course, the possibility existed that he was keeping something back. But that was very unlikely—Maurice just wasn't that type.

The only thing that Francis could do immediately was to seize the water value, which was the easiest to think about. Soon something appeared to make sense, and he began scribbling on the vacant back sheet of a manuscript he had been reading. By then I could not understand what Francis was up to and reverted to *The Times* for amusement. Within a few minutes, however, Francis made me lose all interest in the outside world by telling me that only a small number of formal solutions were compatible both with the Cochran-Crick theory and with Rosy's experimental data. Quickly he began to draw more diagrams to show me how simple the problem was. Though the mathematics eluded me, the crux of the matter was not difficult to follow. Decisions had to be made about the number of polynucleotide chains within the DNA molecule. Superficially, the X-ray data were compatible with two, three, or four strands. It was all a question of the angle and radii at which the DNA strands twisted about the central axis.

By the time the hour-and-a-half train journey was over, Francis saw no reason why we should not know the answer soon. Perhaps a week of solid fiddling with the molecular models would be necessary to make us absolutely sure we had the right answer. Then it would be obvious to the world that Pauling was not the only one capable of true insight into how biological molecules were constructed. Linus' capture of the α-helix was most embarrassing for the Cambridge group. About a year before that triumph, Bragg, Kendrew, and Perutz had published a systematic paper on the conformation of the polypeptide chain, an attack that missed the point. Bragg in fact was still bothered by the fiasco. It hurt his pride at a tender point. There had been previous encounters with Pauling, stretching over a twenty-five-year interval. All too often Linus had got there first.

Even Francis was somewhat humiliated by the event. He was already in the Cavendish when Bragg had become keen about how a polypeptide chain folded up. Moreover, he was privy to a discussion in which the fundamental blunder about the shape of the peptide bond was made. That had certainly been the occasion to interject his critical facility in assessing the meaning of experimental observations—but he had said nothing useful. It was not that Francis normally refrained from criticizing his friends. In other instances he had been annoyingly candid in pointing out where Perutz and Bragg had publicly overinterpreted their hemoglobin results. This open criticism was certainly behind Sir Lawrence's recent outburst against him. In Bragg's view, all that Crick did was to rock the boat.

Now, however, was not the time to concentrate on past mistakes. Instead, the speed with which we talked about possible types of DNA structures gathered intensity as the morning went by. No matter in whose company we found ourselves, Francis would quickly survey the progress of the past few hours, bringing our listener up to date on how we had decided upon models in which the sugar-phosphate backbone was in the center of the molecule. Only in that way would it be possible to obtain a structure regular enough to give the crystalline diffraction patterns observed by Maurice and Rosy. True, we had yet to deal with the irregular sequence of the bases that faced the outside—but this difficulty might vanish in the wash when the correct internal arrangement was located.

There was also the problem of what neutralized the negative charges of the phosphate groups of the DNA backbone. Francis, as well as I, knew almost nothing about how inorganic ions were arranged in three dimensions. We had to face the bleak situation that the world authority on the structural chemistry of ions was Linus Pauling himself. Thus if the crux of the problem was to deduce an unusually clever arrangement of inorganic ions and phosphate groups, we were clearly at a disadvantage. By midday it became imperative to locate a copy of Pauling's classic book, *The Nature of the Chemical Bond*. Then we were having lunch near High Street. Wasting no time over coffee, we dashed into several bookstores until success came in Blackwell's. A rapid reading was made of the relevant sections. This produced the correct values for the exact sizes of the candidate inorganic ions, but nothing that could help push the problem over the top.

When we reached Dorothy's lab in the University Museum, the manic phase had almost passed. Francis ran

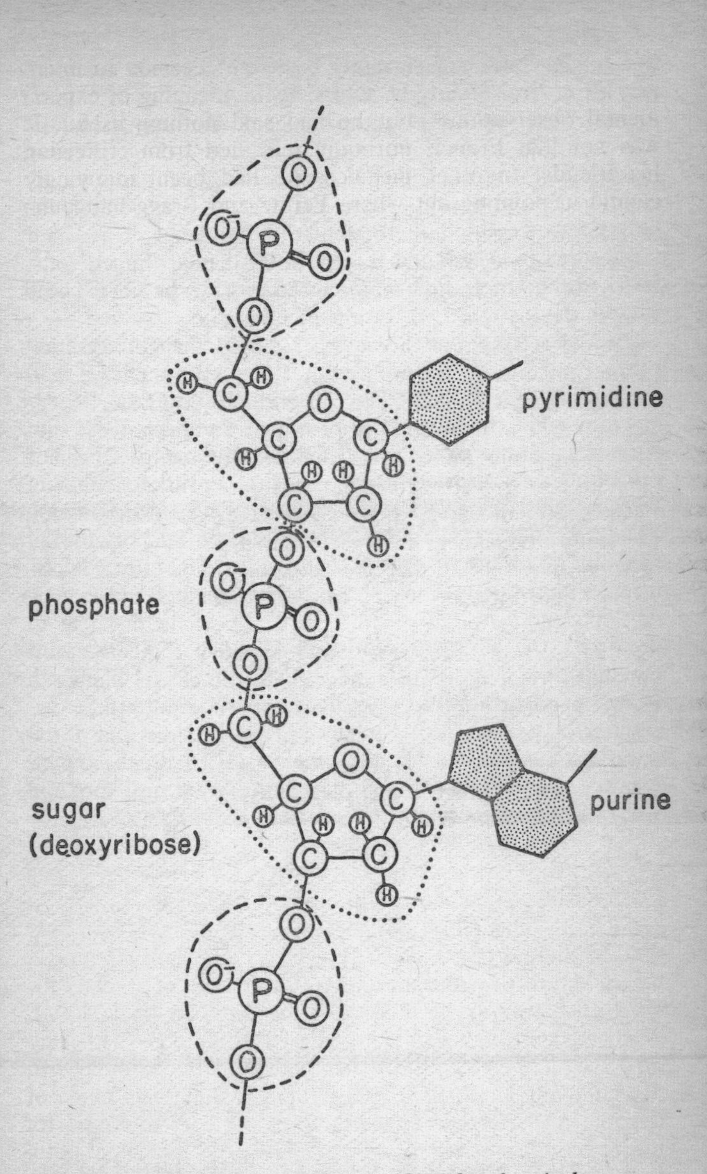

A more detailed view of the covalent bonds of the sugar-phosphate backbone.

through the helical theory itself, devoting only a few minutes to our progress with DNA. Most of the conversation centered instead on Dorothy's recent work with insulin. Since darkness was coming on, there seemed no point in wasting more of her time. We then moved on to Magdalen, where we were to have tea with Avrion Mitchison and Leslie Orgel, both then fellows of the college. Over cakes Francis was ready to talk about trivial things, while I quietly thought how splendid it would be if I could someday live in the style of a Magdalen don.

Dinner with claret, however, restored the conversation to our impending triumph with DNA. By then we had been joined by Francis' close friend, the logician George Kreisel, whose unwashed appearance and idiom did not fit into my picture of the English philosopher. Francis greeted his arrival with great gusto, and the sound of Francis' laughter and Kreisel's Austrian accent dominated the spiffy atmosphere of the restaurant along High Street at which Kreisel had directed us to meet him. For a while Kreisel held forth on a way to make a financial killing by shifting money between the politically divided parts of Europe. Avrion Mitchison then rejoined us, and the conversation for a short time reverted to the casual banter of the intellectual middle class. This sort of small talk, however, was not Kreisel's meat, and so Avrion and I excused ourselves to walk along the medieval streets toward my lodgings. By then I was pleasantly drunk and spoke at length of what we could do when we had DNA.

12

I GAVE John and Elizabeth Kendrew the scoop about DNA when I joined them for breakfast on Monday morning. Elizabeth appeared delighted that success was almost within our grasp, while John took the news more calmly. When it came out that Francis was again in an inspired mood and I had nothing more solid to report than enthusiasm, he became lost to the sections of *The Times* which spoke about the first days of the new Tory government. Soon afterward, John went off to his rooms in Peterhouse, leaving Elizabeth and me to digest the implications of my unanticipated luck. I did not remain long, since the sooner I could get back to the lab, the quicker we could find out which of the several possible answers would be favored by a hard look at the molecular models themselves.

Both Francis and I, however, knew that the models in the Cavendish would not be completely satisfactory. They had been constructed by John some eighteen months before, for the work on the three-dimensional shape of the polypeptide chain. There existed no accurate representations of the groups of atoms unique to DNA. Neither phosphorus atoms nor the purine and pyrimidine bases were on hand. Rapid improvisation would be necessary since there was no time for Max to give a rush order for their construction. Making brand-new models might take all of a week, whereas an answer was possible within a day or so. Thus as soon as I got to the lab I began adding bits of copper wire to some of our carbon-atom models, thereby changing them into the larger-sized phosphorus atoms.

Much more difficulty came from the necessity to fabricate representations of the inorganic ions. Unlike the other constituents, they obeyed no simple-minded rules telling us the angles at which they would form their respective chemical bonds. Most likely we had to know the

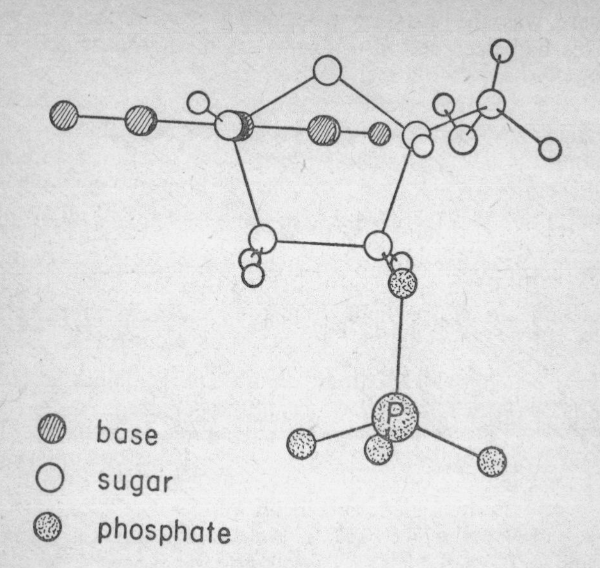

A schematic view of a nucleotide, showing that the plane of the base is almost perpendicular to the plane in which most of the sugar atoms lie. This important fact was established in 1949 by S. Furberg, then working in London at J. D. Bernal's Birkbeck College lab. Later he built some very tentative models for DNA. But not knowing the details of the King's College experiments, he built only single-stranded structures, and so his structural ideas were never seriously considered in the Cavendish.

correct DNA structure before the right models could be made. I maintained the hope, however, that Francis might already be on to the vital trick and would immediately blurt it out when he got to the lab. Over eighteen hours had passed since our last conversation, and there was little chance that the Sunday papers would have distracted him upon his return to the Green Door.

His tenish entrance, however, did not bring the answer. After Sunday supper he had again run through the dilemma but saw no quick answer. The problem was then put aside for a rapid scanning of a novel on the sexual misjudgments of Cambridge dons. The book had its brief good moments, and even in its most ill-conceived pages there was the question of whether any of their friends' lives had been seriously drawn on in the construction of the plot.

Over morning coffee Francis nonetheless exuded confidence that enough experimental data might already be on hand to determine the outcome. We might be able to start the game with several completely different sets of facts and yet always hit the same final answers. Perhaps the whole problem would fall out just by our concentrating on the prettiest way for a polynucleotide chain to fold up. So while Francis continued thinking about the meaning of the X-ray diagram, I began to assemble the various atomic models into several chains, each several nucleotides in length. Though in nature DNA chains are very long, there was no reason to put together anything massive. As long as we could be sure it was a helix, the assignment of the positions for only a couple of nucleotides automatically generated the arrangement of all the other components.

The routine assembly task was over by one, when Francis and I walked over to the Eagle for our habitual lunch with the chemist Herbert Gutfreund. These days John usually went to Peterhouse, while Max always cycled home. Occasionally John's student Hugh Huxley would join us, but of late he was finding it difficult to enjoy Francis' inquisitive lunchtime attacks. For just prior to my arrival in Cambridge, Hugh's decision to take up the problem of how muscles contract had focused Francis' attention on the unforeseen opportunity that, for twenty years or so, muscle physiologists had been accumulating data without tying them into a self-consistent picture. Francis found it a perfect situation for action. There was no need for him to ferret out the relevant experiments since Hugh had already waded through the undigested mass. Lunch after lunch, the facts were put together to form theories which held for a day or so, until Hugh could convince Francis that a result he would like ascribed to experimental error was as solid as the Rock of Gibraltar. Now the construction of Hugh's X-ray camera was completed, and soon he hoped to get experimental evidence to settle the debatable points. The fun would be all lost if somehow Francis could correctly predict what he was going to find.

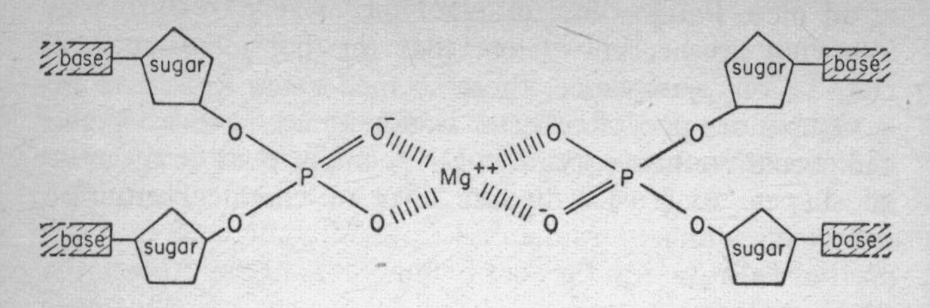

How Mg^{++} ions might be used to bind negatively charged phosphate groups in the center of a compound helix.

But there was no need that day for Hugh to fear a new intellectual invasion. When we walked into the Eagle, Francis did not exchange his usual raucous greetings with the Persian economist Ephraim Eshag, but gave the undistilled impression that something serious was up. The actual model building would start right after lunch, and more concrete plans must be formulated to make the process efficient. So over our gooseberry pie we looked at the pros and cons of one, two, three, and four chains, quickly dismissing one-chain helices as incompatible with the evidence in our hands. As to the forces that held the chains together, the best guess seemed to be salt bridges in which divalent cations like Mg^{++} held together two or more phosphate groups. Admittedly there was no evidence that Rosy's samples contained any divalent ions, and so we might be sticking our necks out. On the other hand, there was absolutely no evidence against our hunch. If only the King's groups had thought about models, they would have asked which salt was present and we would not be placed in this tiresome position. But, with luck, the addition of magnesium or possibly calcium ions to the sugar-phosphate backbone would quickly generate an elegant structure, the correctness of which would not be debatable.

Our first minutes with the models, though, were not joyous. Even though only about fifteen atoms were involved, they kept falling out of the awkward pincers set up to hold them the correct distance from one another. Even worse, the uncomfortable impression arose that there were no obvious restrictions on the bond angles between several of the most important atoms. This was not at all nice. Pauling had cracked the α-helix by ruthlessly following up his knowledge that the peptide bond was flat. To our annoyance, there seemed every reason to believe that the phosphodiester bonds which bound together the successive nucleotides in DNA might exist in a variety of shapes. At least with our level of chemical intuition, there was unlikely to be any single conformation much prettier than the rest.

After tea, however, a shape began to emerge which brought back our spirits. Three chains twisted about each other in a way that gave rise to a crystallographic repeat every 28 Å along the helical axis. This was a feature demanded by Maurice's and Rosy's pictures, so Francis was visibly reassured as he stepped back from the lab bench and surveyed the afternoon's effort. Admittedly a few of the atomic contacts were still too close for comfort, but, after all, the fiddling had just begun. With a few hours' more work, a presentable model should be on display.

Ebullient spirits prevailed during the evening meal at the Green Door. Though Odile could not follow what we were saying, she was obviously cheered by the fact that Francis was about to bring off his second triumph within the month. If this course of events went on, they would soon be rich and could own a car. At no moment did Francis see any point in trying to simplify the matter for Odile's benefit. Ever since she had told him that gravity went only three miles into the sky, this aspect of their relationship was set. Not only did she not know any science, but any attempt to put some in her head would be a losing fight against the years of her convent upbringing. The most to hope for was an appreciation of the linear way in which money was measured.

Our conversation instead centered upon a young art student then about to Marry Odile's friend Harmut Weil. This capture was mildly displeasing to Francis. It was about to remove the prettiest girl from their party circle. Moreover, there was more than one thing cloudy about Harmut. He had come out of a German university tradition that believed in dueling. There was also his undeniable skill in persuading numerous Cambridge women to pose for his camera.

All thought of women, however, was banished by the time Francis breezed into the lab just before morning coffee. Soon, when several atoms had been pushed in or out, the three-chain model began to look quite reasonable. The next obvious step would be to check it with Rosy's quantitative measurements. The model would certainly fit with the general locations of the X-ray reflections, for its essential helical parameters had been chosen to fit the seminar facts I had conveyed to Francis. If it were right, however, the model would also accurately predict the relative intensities of the various X-ray reflections.

A quick phone call was made to Maurice. Francis explained how the helical diffraction theory allowed a rapid survey of possible DNA models, and that he and I had just come up with a creature which might be the answer we were all awaiting. The best thing would be for Maurice immediately to come and look it over. But Maurice gave no definite date, saying he thought he might make it sometime within the week. Soon after the phone was put down, John came in to see how Maurice had taken the news of the breakthrough. Francis found it hard to sum up his reply. It was almost as if Maurice were indifferent to what we were doing.

In the midst of further fiddling that afternoon, a call came through from King's. Maurice would come up on the 10:10 train from London the following morning. Moreover, he would not be alone. His collaborator Willy Seeds would also come. Even more to the point was that Rosy, together with her student R. G. Gosling, would be on the same train. Apparently they were still interested in the answer.

13

MAURICE decided to take a cab from the station to the lab. Ordinarily he would have come by bus, but now there were four of them to share the cost. Moreover, there would be no satisfaction in waiting at the bus stop with Rosy. It would make the present uncomfortable situation worse than it need be. His well-intentioned remarks never came off, and even now, when the possibility of humiliation hung over them, Rosy was as indifferent as ever to his presence and directed all her attention to Gosling. There was only the slightest effort made at a united appearance when Maurice poked his head into our lab to say they had come. Especially in sticky situations like this, Maurice thought that a few minutes without science was the way to proceed. Rosy, however, had not come here to throw out foolish words, but quickly wanted to know where things stood.

Neither Max nor John did anything to take the stage away from Francis. This was his day, and after they came in to greet Maurice they both pleaded pressure of their work to retire behind the closed doors of their joint office. Before the delegation's arrival, Francis and I had agreed to reveal our progress in two stages. Francis would first sum up the advantages of the helical theory. Then together we could explain how we had arrived at the proposed model for DNA. Afterwards we could go to the Eagle for lunch, leaving the afternoon free to discuss how we could all proceed with the final phases of the problem.

The first part of the show ran on schedule. Francis saw no reason to understate the power of the helical theory and within several minutes revealed the way Bessel functions gave neat answers. None of the visitors, however, gave any indication of sharing Francis' delight. Instead of wishing to do something with the pretty equations, Maurice wanted to concentrate on the fact that the theory did not go beyond some mathematics his colleague

Stokes had worked out without all this fanfare. Stokes had solved the problem in the train while going home one evening and had produced the theory on a small sheet of paper the next morning.

Rosy did not give a hoot about the priority of the creation of the helical theory and, as Francis prattled on, she displayed increasing irritation. The sermon was unnecessary, since to her mind there was not a shred of evidence that DNA was helical. Whether this was the case would come out of further X-ray work. Inspection of the model itself only increased her disdain. Nothing in Francis' argument justified all this fuss. She became positively aggressive when we got on the topic of Mg^{++} ions that held together the phosphate groups of our three-chain model. This feature had no appeal at all to Rosy, who curtly pointed out that the Mg^{++} ions would be surrounded by tight shells of water molecules and so were unlikely to be the kingpins of a tight structure.

Most annoyingly, her objections were not mere perversity: at this stage the embarrassing fact came out that my recollection of the water content of Rosy's DNA samples could not be right. The awkward truth became apparent that the correct DNA model must contain at least ten times more water than was found in our model. This did not mean that we were necessarily wrong—with luck the extra water might be fudged into vacant regions on the periphery of our helix. On the other hand, there was no escaping the conclusion that our argument was soft. As soon as the possibility arose that much more water was involved, the number of potential DNA models alarmingly increased.

Though Francis could not help dominating the lunchtime conversation, his mood was no longer that of a confident master lecturing hapless colonial children who until then had never experienced a first-rate intellect. The group holding the ball was clear to all. The best way to salvage something from the day was to come to an agreement about the next round of experiments. In particular, only a few weeks' work should be necessary to see whether the DNA structure was dependent upon the exact ions used to neutralize the negative phosphate groups. Then the beastly uncertainty as to whether Mg^{++} ions were important could vanish. With this accomplished, another round of model building could start and, given luck, it might occur by Christmas.

Our subsequent after-lunch walk into King's and along the backs to Trinity did not, however, reveal any converts. Rosy and Gosling were pugnaciously assertive: their future course of action would be unaffected by their fifty-mile excursion into adolescent blather. Maurice and Willy Seeds gave more indication of being reasonable, but there was no certainty that this was anything more than a reflection of a desire not to agree with Rosy.

The situation did not improve when we got back to the lab. Francis did not want to surrender immediately, so he went through some of the actual details of how we went about the model building. Nonetheless, he quickly lost heart when it became apparent that I was the only one joining the conversation. Moreover, by this time neither of us really wanted to look at our model. All its glamor had vanished, and the crudely improvised phosphorus atoms gave no hint that they would ever neatly fit into something of value. Then when Maurice mentioned that, if they moved with haste, the bus might enable them to get the 3:40 train to Liverpool Street Station, we quickly said good-bye.

14

Rosy's triumph all too soon filtered up the stairs to Bragg. There was nothing to do but appear unperturbed as the news of the upset confirmed the fact that Francis might move faster if occasionally he would close his mouth. The consequences spread in a predictable fashion. Clearly this was the moment for Maurice's boss to discuss with Bragg whether it made sense for Crick and the American to duplicate King's heavy investment in DNA.

Sir Lawrence had had too much of Francis to be surprised that he had again stirred up an unnecessary tempest. There was no telling where he would let loose the next explosion. If he continued to behave this way, he could easily spend the next five years in the lab without collecting sufficient data to warrant an honest Ph.D. The chilling prospect of enduring Francis throughout the remaining years of his tenure as the Cavendish Professor was too much to ask of Bragg or anyone with a normal set of nerves. Besides, for too long he had lived under the shadow of his famous father, with most people falsely thinking that his father, not he, was responsible for the sharp insight behind Bragg's Law. Now when he should be enjoying the rewards accorded the most prestigious chair in science, he had to be responsible for the outrageous antics of an unsuccessful genius.

The decision was thus passed on to Max that Francis and I must give up DNA. Bragg felt no qualms that this might impede science, since inquiries to Max and John had revealed nothing original in our approach. After Pauling's success, no one could claim that faith in helices implied anything but an uncomplicated brain. Letting the King's group have the first go at helical models was the right thing in any circumstance. Crick could then buckle down to his thesis task of investigating the ways that hemoglobin crystals shrink when they are placed in salt solutions of different density. A year to eighteen months of steady work might tell something more solid about the shape of the hemoglobin molecule. With a Ph.D. in his pocket Crick could then seek employment elsewhere.

No attempt was made to appeal the verdict. To the relief of Max and John, we refrained from publicly questioning Bragg's decision. An open outcry would reveal that our professor was completely in the dark about what the initials DNA stood for. There was no reason to believe that he gave it one hundredth the importance of the structure of metals, for which he took great delight in making soap-bubble models. Nothing then gave Sir Lawrence more pleasure than showing his ingenious motion-picture film of how bubbles bump each other.

Our reasonableness did not arise, however, from a desire to keep peace with Bragg. Lying low made sense because we were up the creek with models based on sugar-phosphate cores. No matter how we looked at them, they smelled bad. On the day following the visit from King's, a hard look was given both to the ill-fated three-chain affair and to a number of possible variants. One couldn't be sure, but the impression was there that any model placing the sugar-phosphate backbone in the center of a helix forced atoms closer together than the laws of chemistry allowed. Positioning one atom the proper distance from its neighbor often caused a distant atom to become jammed impossibly close to its partners.

A fresh start would be necessary to get the problem rolling again. Sadly, however, we realized that the impetuous tangle with King's would dry up our source of new experimental results. Subsequent invitations to the research colloquia were not to be expected, and even the most casual questioning of Maurice would provoke the suspicion that we were at it again. What was worse was the virtual certainty that cessation of model building on our part would not be accompanied by a burst of corresponding activity in their lab. So far, to our knowledge, King's had not built any three-dimensional models of the necessary atoms. Nonetheless, our offer to speed that task by giving them the Cambridge molds for making the models was only halfheartedly received. Maurice did say, though, that within a few weeks someone might be found to put something together, and it was arranged that the next time one of us went down to London the jigs could be dropped off at their lab.

Thus the prospects that anyone on the British side of the Atlantic would crack DNA looked dim as the Christmas holidays drew near. Though Francis went back to proteins, obliging Bragg by working on his thesis was not to his liking. Instead, after a few days of relative silence, he began to spout about superhelical arrangements of the α-helix itself. Only during the lunch hour could I be sure that he would talk DNA. Fortunately, John Kendrew sensed that the moratorium on working on DNA did not extend to thinking about it. At no time did he try to reinterest me in myoglobin. Instead, I used the dark and chilly days to learn more theoretical chemistry or to leaf through journals, hoping that possibly there existed a forgotten clue to DNA.

The book I poked open the most was Francis' copy of *The Nature of the Chemical Bond*. Increasingly often, when Francis needed it to look up a crucial bond length, it would turn up on the quarter bench of lab space that John had given to me for experimental work. Somewhere in Pauling's masterpiece I hoped the real secret would lie. Thus Francis' gift to me of a second copy was a good omen. On the flyleaf was the inscription, "To Jim from Francis—Christmas '51." The remnants of Christianity were indeed useful.

15

I DID not sit through the Christmas holidays in Cambridge. Avrion Mitchison had invited me to Carradale, the home of his parents, on the Mull of Kintyre. This was real luck, since over holidays Av's mother, Naomi, the distinguished writer, and his Labor MP father, Dick, were known to fill their large house with odd assortments of lively minds. Moreover, Naomi was a sister of England's most clever and eccentric biologist, J. B. S. Haldane. Neither the feeling that our DNA work had hit a roadblock nor the uncertainty of getting paid for the year was of much concern when I joined Av and his sister Val at Euston Station. No seats were left on the overnight Glasgow train, giving us a ten-hour journey seated on luggage listening to Val comment on the dull, boorish habits of the Americans who each year are deposited in increasing numbers at Oxford.

At Glasgow we found my sister Elizabeth, who had flown to Prestwick from Copenhagen. Two weeks previously she had sent a letter relating that she was pursued by a Dane. Instantly I sensed impending disaster, for he was a successful actor. At once I inquired whether I could bring Elizabeth to Carradale. The affirmative reply I received with much relief, since it was inconceivable that my sister could think about settling in Denmark after two weeks of an eccentric country house.

Dick Mitchison met the Campbelltown bus at the turnoff for Carradale to drive us the final twenty hilly miles to the tiny Scottish fishing village where he and Naomi had lived for the past twenty years. Dinner was still going on as we emerged from a stone passage, which connected the gunroom with several larders, into a dining room dominated by sharp authoritative chatter. Av's zoologist brother Murdoch had already come, and he enjoyed cornering people to talk about how cells divide. More often, the theme was politics and the awkward cold war thought up by the American paranoids, who should be back in the law offices of middlewestern towns.

By the following morning I was aware that the best way not to feel impossibly cold was to remain in bed or, when that proved impossible, to go walking, unless the rain was coming down in buckets. In the afternoons Dick was always trying to get someone to shoot pigeons, but after one attempt, when I fired the gun after the pigeons were out of view, I took to lying on the drawing-room floor as close as possible to the fire. There was also the warming diversion of going to the library to play ping-pong beneath Wyndham Lewis' stern drawings of Naomi and her children.

More than a week passed before I slowly caught on that a family of leftish leanings could be bothered by the way their guests dressed for dinner, but I put this aberrant behavior down as a sign of approaching old age. The thought never occurred to me that my own appearance was noticed, since my hair was beginning to lose its American identity. Odile had been very shocked when Max introduced me to her on my first day in Cambridge and afterwards had told Francis that a bald American was coming to work in the lab. The best way to rectify the situation was to avoid a barber until I merged with the Cambridge scene. Though my sister was upset when she saw me, I knew that months, if not years, might be required to replace her superficial values with those of the English intellectual. Carradale thus was the perfect environment to go one step further and acquire a beard. Admittedly I did not like its reddish color, but shaving with cold water was agony. Yet after a week of Val's and Murdoch's acid comments, together with the expected unpleasantness of my sister, I emerged for dinner with a clean face. When Naomi made a complimentary remark about my looks, I knew that I had made the right decision.

In the evenings there was no way to avoid intellectual games, which gave the greatest advantage to a large vocabulary. Every time my limpid contribution was read, I wanted to sink behind my chair rather than face the condescending stares of the Mitchison women. To my relief, the large number of house guests never permitted my turn to come often, and I made a point of sitting near the eve-

Francis Crick and J. D. Watson during a walk along the backs. In the distance, King's College Chapel.

Francis next to a Cavendish X-ray tube.

Maurice Wilkins.

Snapshot taken at the microbial genetics meeting, held at the Institute for Theoretical Physics, Copenhagen, March 1951. First row: O. Maaløe, R. Latarjet, E. Wollman. Second row: N. Bohr, N. Visconti, G. Ehrensvaard, W. Weidel, H. Hyden, V. Bonifas, G. Stent, H. Kalckar, B. Wright, J. D. Watson, M. Westergaard.

Linus Pauling with his atomic models.

Sir Lawrence Bragg sitting at his Cavendish desk.

Rosalind Franklin.

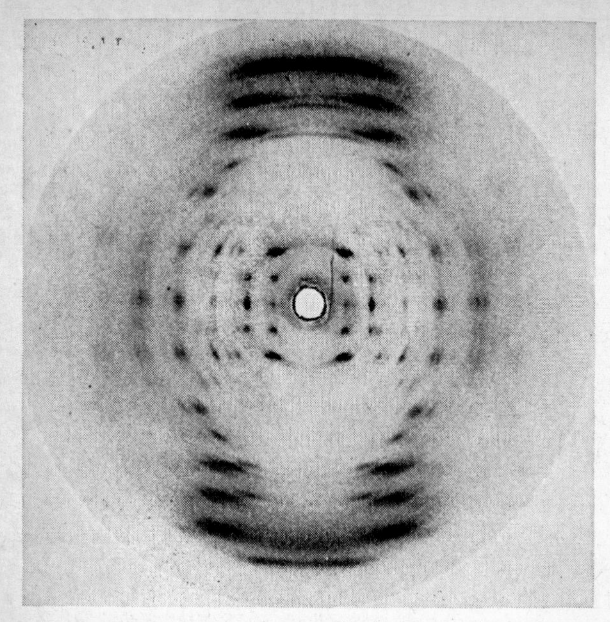

An X-ray photograph of crystalline DNA in the A form.

Elizabeth Watson, with Clare Bridge in the background.

In Paris on the way to the Riviera, spring 1952.

The meeting at Royaumont, July 1952.

Vacation in the Italian Alps, August 1952.

November 1952

A Hypothetical Scheme of the Interrelationship between the Nucleic Acids and Proteins

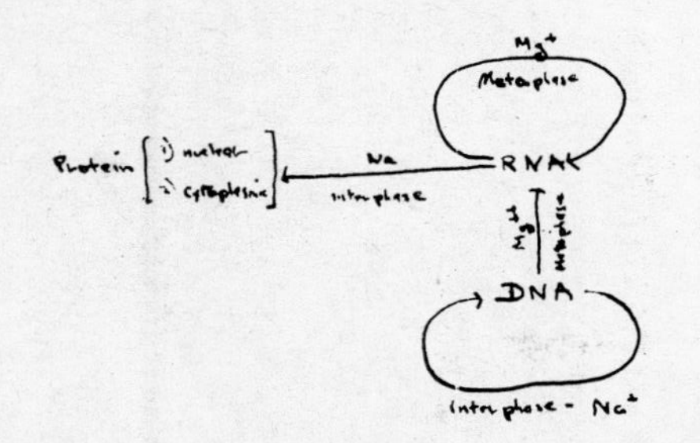

Consequences of scheme

1. RNA synthesis and DNA synthesis should not occur at the same time. Protein synthesis and DNA synthesis will occur simultaneously.
2. Nuclear RNA synthesis will occur only in dividing cells.
3. The [illegible] Mg^{++} concentration will increase toward metaphase and decrease during interphase
4. The content of nucleolar RNA may possibly remain constant during interphase. - Synthesis of nucleolar RNA occurs on the chromosomes during prophase - metaphase

Early ideas on the DNA-RNA-protein relation.

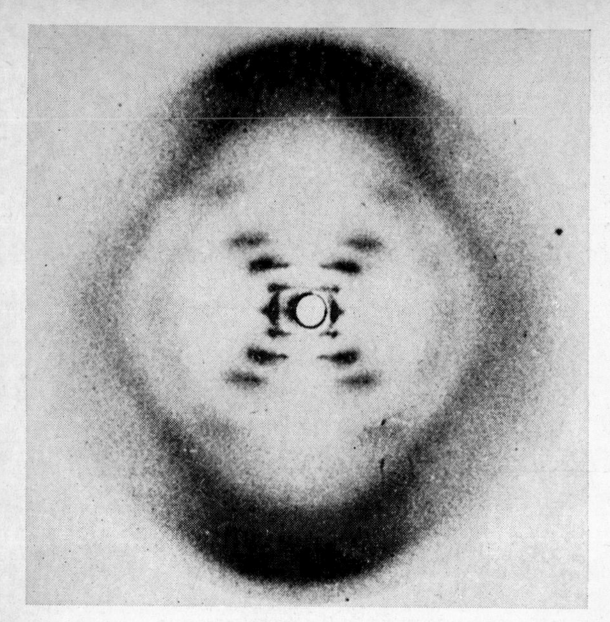

An X-ray photograph of DNA in the B form, taken by Rosalind Franklin late in 1952.

Watson and Crick in front of the DNA model.

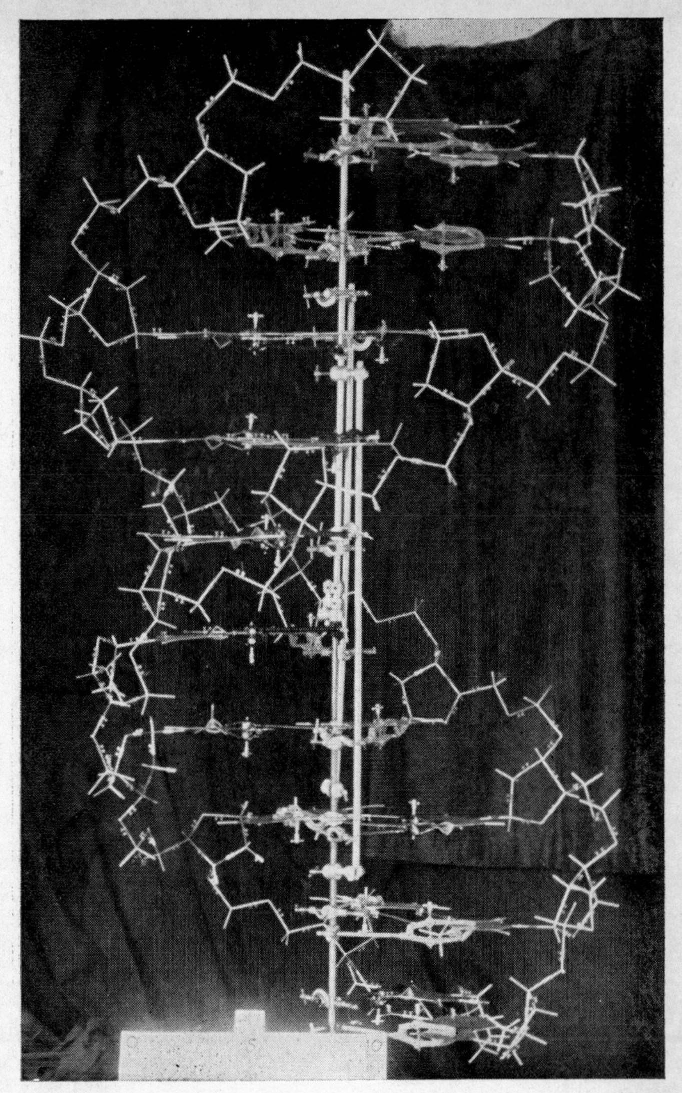

The original demonstration model of the double helix (the scale gives distances in Angstroms).

On the six following pages: The letter written to Delbrück telling of the double helix.

UNIVERSITY OF CAMBRIDGE DEPARTMENT OF PHYSICS

TELEPHONE
CAMBRIDGE 55478

CAVENDISH LABORATORY
FREE SCHOOL LANE
CAMBRIDGE

March 12, 1953

Dear Max

Thank you very much for your recent letters. We were quite interested in your account of the Pauling Seminar. The day following the arrival of your letter, I received a note from Pauling, mentioning that their model had been revised, and indicating interest in our model. We shall thus have to write him in the near future as to what we are doing. Until now we preferred not to write him since we did not want to commit ourselves until we were completely sure that all of the van der Waals contacts were correct and that all aspects of our structure were stereochemically feasible. I believe now that we have made sure that our structure can be built, and today we are laboriously calculating out exact atomic coordinates.

Our model (a joint project of Francis Crick and myself) bears no relationship to either the original or to the revised Pauling-Corey-Schomaker models. It is a strange model and embodies several unusual features. However since DNA is an unusual substance we are not hesitant in being bold. The main features of the model are (1) The basic structure is helical – it consists of two intertwining helices – the core of the helix is occupied by the purine and pyrimidine bases. – The phosphate groups are on the outside (2) The helices are not identical but complementary so that if one helix contains a purine base, the other helix contains a pyrimidine – this feature is a result of our attempt to make the residues equivalent and at the same time put the purines and pyrimidine bases in the center. The pairing of the purine with pyrimidine is very exact and dictated by their desire to form hydrogen bonds – Adenine will pair with Thymine while Guanine will always pair with Cytosine. For example

Adenine — sugar

(next page)

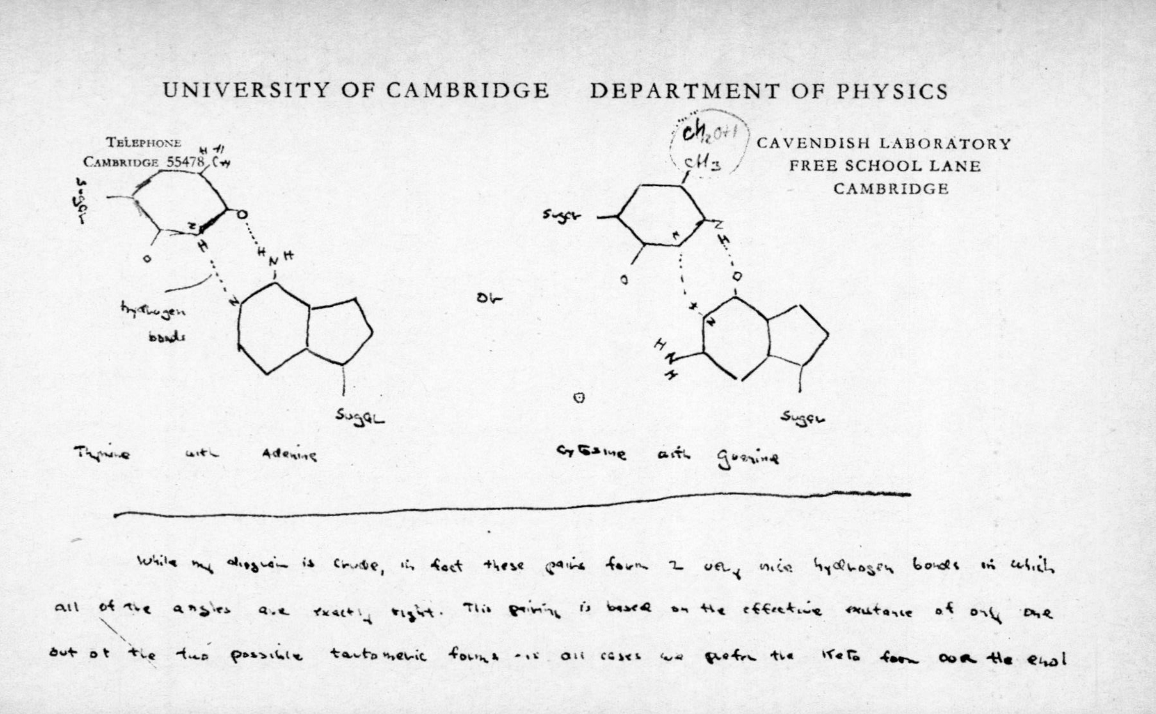
UNIVERSITY OF CAMBRIDGE DEPARTMENT OF PHYSICS

TELEPHONE
CAMBRIDGE 55478

CAVENDISH LABORATORY
FREE SCHOOL LANE
CAMBRIDGE

While my diagram is crude, in fact these pairs form 2 very nice hydrogen bonds in which all of the angles are exactly right. This pairing is based on the effective existance of only one out of the two possible tautomeric forms – in all cases we prefer the keto form over the enol are the amino over the imino. This is a definitely an assumption but Jerry Donohue and Bill Cochran tell us that, for all organic molecules so far examined, the keto and amino forms are present in preference to the enol and imino possibilities.

The model has been derived almost entirely from stereochemical considerations with the only X-ray consideration being the spacing between the pair of bases 3.4 A which was originally found by Astbury. It tends to build itself with approximately 10 residues per turn in 34 A. The screw is right handed.

The X-ray pattern approximately agrees with the model, but since the photographs available to us are poor and meagre (we have no photographs of our own and like Pauling must use Astbury's photographs) this agreement in no way constitutes a proof of our model. We are certainly a long way from proving its correctness. To do this we must obtain collaboration from the group at Kings College London who possess very excellent photographs of a crystalline phase in addition to rather good photographs of a paracrystalline phase. Our model has been made in reference to the paracrystalline form and as yet we have no clear ideas as to how these helices can

UNIVERSITY OF CAMBRIDGE DEPARTMENT OF PHYSICS

TELEPHONE
CAMBRIDGE 55478

CAVENDISH LABORATORY
FREE SCHOOL LANE
CAMBRIDGE

pack together to form the crystalline phase.

In the next day or so Crick and I shall send a note to Nature proposing our structure as a possible model, at the same time emphasizing its provisional nature and the lack of proof in its favor. Even if wrong I believe it to be interesting since it provides a concrete example of a structure composed of complementary chains. If by chance, it is right then I suspect we may be making a slight dent into the manner in which DNA can reproduce itself. For these reasons (in addition to many others) I prefer this type of model over Pauling's which if true would tell us next to nothing about manner of DNA reproduction.

I shall write you in a day or so about the recombination paper. Yesterday I received a very interesting note from Bill Hayes. I believe he is sending you a copy.

I have met Alfred Tissieres recently. He seems very nice. He speaks fondly of Pasadena and I suspect has not yet become accustomed to being a Fellow of Kings.

My regards to Henry

Jim

P.S. We would prefer your not mentioning this letter to Pauling. When our letter to Nature is completed we shall send him a copy. We should like to send him coordinates.

Morning coffee in the Cavendish just after publication of the manuscript on the double helix.

In Stockholm for their Nobel Prizes, December 1962: Maurice Wilkins, John Steinbeck, John Kendrew, Max Perutz, Francis Crick, and James D. Watson.

ning's box of chocolates, hoping no one would notice that I never passed it. Much more agreeable were the hours playing "Murder" in the dark twisting recesses of the upstairs floors. The most ruthless of the murder addicts was Av's sister Lois, then just back from teaching for a year in Karachi, and a firm proponent of the hypocrisy of Indian vegetable eaters.

Almost from the start of my stay I knew that I would depart from Naomi's and Dick's spectrum of the left with the greatest reluctance. The prospect of lunch with the alcoholic English cider more than compensated for the habit of leaving the outside doors open to the westerly winds. My departure, three days after the New Year, nonetheless had been fixed by Murdoch's arranging for me to speak at a London meeting of the Society for Experimental Biology. Two days before my scheduled departure there was a heavy fall of snow, giving to the barren moors the look of Antarctic mountains. It was a perfect occasion for a long afternoon walk along the closed Campbelltown Road, with Av talking about his thesis experiments on the transplantation of immunity while I thought about the possibility that the road might remain impassable through the day I was to leave. The climate was not with me, however, for a group from the house caught the Clyde steamer at Tarbert and the next morning we were in London.

Upon my return to Cambridge I had expected to hear from the States about my fellowship, but there was no official communication to greet me. Since Luria had written me in November not to worry, the absence of firm news by now seemed ominous. Apparently no decision had been made and the worst was to be expected. The ax, however, could at most be only annoying. John and Max gave me assurance that a small English stipend could be dug up if I was completely cut off. Only in late January did my suspense end, with the arrival of a letter from Washington: I was sacked. The letter quoted the section of the fellowship award stating that the fellowship was valid only for work in the designated institution. My violation of this provision gave them no choice but to revoke the award.

The second paragraph gave the news that I had been awarded a completely new fellowship. I was not, however, to be let off merely with the long period of uncertainty. The second fellowship was not for the customary twelve-month period but explicitly terminated after eight months, in the middle of May. My real punishment in not following the Board's advice and going to Stockholm was a thousand dollars. By this time it was virtually impossible to obtain any support which could begin before the September start of a new school year. I naturally accepted the fellowship. Two thousand dollars was not to be thrown away.

Less than a week later, a new letter came from Washington. It was signed by the same man, but not as head of the fellowship board. The hat he now displayed was that of the chairman of a committee of the National Research Council. A meeting was being arranged for which I was asked to give a lecture on the growth of viruses. The time of the meeting, to be held in Williamstown, was the middle of June, only a month after my fellowship would expire. I, of course, had not the slightest intention of leaving either in June or in September. The only problem was how to frame an answer. My first impulse was to write that I could not come because of an unforeseen financial disaster. But on second thought, I was against giving him the satisfaction of thinking he had affected my affairs. A letter went off saying that I found Cambridge intellectually very exciting and so did not plan to be in the States by June.

16

By now I had decided to mark time by working on tobacco mosaic virus (TMV). A vital component of TMV was nucleic acid, and so it was the perfect front to mask my continued interest in DNA. Admittedly the nucleic-acid component was not DNA but a second form of nucleic acid known as ribonucleic acid (RNA). The difference was an advantage, however, since Maurice could lay no claim to RNA. If we solved RNA we might also provide the vital clue to DNA. On the other hand, TMV was thought to have a molecular weight of 40 million and at first glance should be frightfully more difficult to understand than the much smaller myoglobin and hemoglobin molecules that John and Max had been working on for years without obtaining any biologically interesting answers.

Moreover, TMV had previously been looked at with X rays by J. D. Bernal and I. Fankuchen. This in itself was scary, since the scope of Bernal's brain was legendary and I could never hope to have his grasp of crystallographic theory. I was even unable to understand large sections of their classic paper published just after the start of the war in the *Journal of General Physiology*. This was an odd place to publish, but Bernal had become absorbed in the war effort, and Fankuchen, by then returned to the States, decided to place their data in a journal looked at by people interested in viruses. After the war Fankuchen lost interest in viruses, and, though Bernal dabbled at protein crystallography, he was more concerned about furthering good relations with the Communist countries.

Though the theoretical basis for many of their conclusions was shaky, the take-home lesson was obvious. TMV was constructed from a large number of identical subunits. How the subunits were arranged they did not know. Moreover, 1939 was too early to come to grips with the fact that the protein and RNA components were likely to be constructed along radically different lines. By now, however, protein subunits were easy to imagine in large numbers. Just the opposite was true of RNA. Division of the RNA component into a large number of subunits would produce polynucleotide chains too small to carry the genetic information that Francis and I believed must reside in the viral RNA. The most plausible hypothesis for the TMV structure was a central RNA core surrounded by a large number of identical small protein subunits.

In fact, there already existed biochemical evidence for protein building blocks. Experiments of the German Gerhard Schramm, first published in 1944, reported that TMV particles in mild alkali fell apart into free RNA and a large number of similar, if not identical, protein molecules. Virtually no one outside Germany, however, thought that Schramm's story was right. This was because of the war. It was inconceivable to most people that the German beasts would have permitted the extensive experiments underlying his claims to be routinely carried out during the last years of a war they were so badly losing. It was all too easy to imagine that the work had direct Nazi support and that his experiments were incorrectly analyzed. Wasting time to disprove Schramm was not to most biochemists' liking. As I read Bernal's paper, however, I suddenly became enthusiastic about Schramm, for, if he had misinterpreted his data, by accident he had hit upon the right answer.

Conceivably a few additional X-ray pictures would tell how the protein subunits were arranged. This was particularly true if they were helically stacked. Excitedly I pilfered Bernal's and Fankuchen's paper from the Philosophical Library and brought it up to the lab so that Francis could inspect the TMV X-ray picture. When he saw the blank regions that characterize helical patterns, he jumped into action, quickly spilling out several possible helical TMV structures. From this moment on, I knew I could no longer avoid actually understanding the helical theory. Waiting until Francis had free time to help me would save me from having to master the mathematics, but only at the penalty of my standing still if Francis was out of the room. Luckily, merely a superficial grasp was needed to see why the TMV X-ray picture suggested a helix with a turn every 23 Å along the helical axis. The rules were, in fact, so simple that Francis considered writing them up under the title, "Fourier Transforms for the Birdwatcher."

This time, however, Francis did not carry the ball and on subsequent days maintained that the evidence for a TMV helix was only so-so. My morale automatically went down, until I hit upon a foolproof reason why subunits should be helically arranged. In a moment of after-supper boredom I had read a Faraday Society Discussion on "The Structure of Metals." It contained an ingenious theory by the theoretician F. C. Frank on how crystals grow. Every time the calculations were properly done, the paradoxical answer emerged that the crystals could not grow at anywhere near the observed rates. Frank saw that the paradox vanished if crystals were not as regular as suspected, but contained dislocations resulting in the perpetual presence of cozy corners into which new molecules could fit.

Several days later, on the bus to Oxford, the notion came to me that each TMV particle should be thought of as a tiny crystal growing like other crystals through the possession of cozy corners. Most important, the simplest way to generate cozy corners was to have the subunits helically arranged. The idea was so simple that it had to be right. Every helical staircase I saw that weekend in Oxford made me more confident that other biological structures would also have helical symmetry. For over a week I pored over electron micrographs of muscle and collagen fibers, looking for hints of helices. Francis, however, remained lukewarm, and in the absence of any hard facts I knew it was futile to try to bring him around.

Hugh Huxley came to my rescue by offering to teach me how to set up the X-ray camera for photographing TMV. The way to reveal a helix was to tilt the oriented TMV sample at several angles to the X-ray beam. Fankuchen had not done this, since before the war no one took helices seriously. I thus went to Roy Markham to see if any spare TMV was on hand. Markham then worked in the Molteno Institute, which, unlike all other Cambridge labs, was well heated. This unusual state came from the asthma of David Keilin, then the "Quick Professor" and Director of Molteno. I always welcomed an excuse to exist momentarily at 70° F, even though I was never sure when Markham would start the conversation by saying how bad I looked, implying that if I had been brought up on English beer I would not be in my sorry state. This time he was unexpectedly sympathetic and without hesitation volunteered some virus. The idea of Francis and me dirtying our hands with experiments brought unconcealed amusement.

My first X-ray pictures revealed, not unexpectedly, much less detail than was found in the published pictures. Over a month was required before I could get even half-way presentable pictures. They were still a long way, though, from being good enough to spot a helix. The only real fun during February came from a costume party given by Geoffrey Roughton at his parents' home on Adams Road. Surprisingly, Francis did not wish to go, even though Geoffrey knew many pretty girls and was said to write poetry wearing one earring. Odile, however, did not want to miss it, so I went with her after hiring a Restoration soldier's garb. The moment we edged through the door into the crush of half-drunken dancers we knew the evening would be a smashing success, since seemingly half the attractive Cambridge *au pair* girls (foreign girls living with English families) were there.

A week later there was a Tropical Night Ball that Odile was keen to attend, both since she had done the decorations and because it was sponsored by black people. Francis again demurred, this time wisely. The dance floor was half vacant, and even after several long drinks I did not enjoy dancing badly in open view. More to the point was that Linus Pauling was coming to London in May for a meeting organized by the Royal Society on the structure of proteins. One could never be sure where he would strike next. Particularly chilling was the prospect that he would ask to visit King's.

17

LINUS, however, was blocked from descending on London. His trip abruptly terminated at Idlewild through the removal of his passport. The State Department did not want troublemakers like Pauling wandering about the globe saying nasty things about the politics of its onetime investment bankers who held back the hordes of godless Reds. Failure to contain Pauling might result in a London press conference with Linus expounding peaceful coexistence. Acheson's position was harassed enough without giving McCarthy the opportunity to announce that our government let radicals protected by U.S. passports set back the American way of life.

Francis and I were already in London when the scandal reached the Royal Society. The reaction was one of almost complete disbelief. It was far more reassuring to go on imagining that Linus had taken ill on the plane to New York. The failure to let one of the world's leading scientists attend a completely nonpolitical meeting would have been expected from the Russians. A first-rate Russian might easily abscond to the more affluent West. No danger existed, however, that Linus might want to flee. Only complete satisfaction with their Cal Tech existence came from him and his family.

Several members of Cal Tech's governing board, however, would have been delighted with his voluntary departure. Every time they picked up a newspaper and saw Pauling's name among the sponsors of the World Peace Conference they seethed with rage, wishing there were a way to rid Southern California of his pernicious charm. But Linus knew better than to expect more than confused anger from the self-made California millionaires whose knowledge of foreign policy was formed largely by the *Los Angeles Times*.

The debacle was no surprise to several of us who had just been in Oxford for a Society of General Microbiology meeting on "The Nature of Viral Multiplication." One of the main speakers was to have been Luria. Two weeks prior to his scheduled flight to London, he was notified that he would not get a passport. As usual, the State Department would not come clean about what it considered dirt.

Luria's absence thrust upon me the job of describing the recent experiments of the American phage workers. There was no need to put together a speech. Several days before the meeting, Al Hershey had sent me a long letter from Cold Spring Harbor summarizing the recently completed experiments by which he and Martha Chase established that a key feature of the infection of a bacterium by a phage was the injection of the viral DNA into the host bacterium. Most important, very little protein entered the bacterium. Their experiment was thus a powerful new proof that DNA is the primary genetic material.

Nonetheless, almost no one in the audience of over four hundred microbiologists seemed interested as I read long sections of Hershey's letter. Obvious exceptions were André Lwoff, Seymour Benzer, and Gunther Stent, all briefly over from Paris. They knew that Hershey's experiments were not trivial and that from then on everyone was going to place more emphasis on DNA. To most of the spectators, however, Hershey's name carried no weight. Moreover, when it came out that I was an American, my uncut hair provided no assurance that my scientific judgment was not equally bizarre.

Dominating the meeting were the English plant virologists F. C. Bawden and N. W. Pirie. No one could match the smooth erudition of Bawden or the assured nihilism of Pirie, who strongly disliked the notion that some phages have tails or that TMV is of fixed length. When I tried to corner Pirie about Schramm's experiments he said they should be dismissed, and so I retreated to the politically less controversial point of whether the 3000 Å length of many TMV particles was biologically important. The idea that a simple answer was preferable had no appeal to Pirie, who knew that viruses were too large to have well-defined structures.

If it had not been for the presence of Lwoff, the meeting would have flopped totally. André was very keen about the role of divalent metals in phage multiplication and so was receptive to my belief that ions were decisively important for nucleic-acid structure. Especially intriguing was his hunch that specific ions might be the trick for the exact copying of macromolecules or the attraction between similar chromosomes. There was no way to test our dreams, however, unless Rosy did an about-face from her determination to rely completely on classical X-ray diffraction techniques.

At the Royal Society Meeting there was no hint that anyone at King's had mentioned ions since the confrontation with Francis and me in early December. Upon pressing Maurice, I learned that the jigs for the molecular models had not been touched after arriving at his lab. The time had not yet come to press Rosy and Gosling about building models. If anything, the squabbling between Maurice and Rosy was more bitter than before the visit to Cambridge. Now she was insisting that her data told her DNA was *not* a helix. Rather than build helical models at Maurice's command, she might twist the copper-wire models about his neck.

When Maurice asked whether we needed the molds back in Cambridge, we said yes, half implying that more carbon atoms were needed to make models showing how polypeptide chains turned corners. To my relief, Maurice was very open about what was not happening at King's. The fact that I was doing serious X-ray work with TMV gave him assurance that I should not soon again become preoccupied with the DNA pattern.

18

MAURICE had no suspicion that almost immediately I would get the X-ray pattern needed to prove that TMV was helical. My unexpected success came from using a powerful rotating anode X-ray tube which had just been assembled in the Cavendish. This supertube permitted me to take pictures twenty times faster than with conventional equipment. Within a week I more than doubled the number of my TMV photographs.

Custom then locked the doors of the Cavendish at 10:00 P.M. Though the porter had a flat next to the gate, no one disturbed him after the closing hour. Rutherford had believed in discouraging students from night work, since the summer evenings were more suitable for tennis. Even fifteen years after his death there was only one key available for late workers. This was now pre-empted by Hugh Huxley, who argued that muscle fibers were living and hence not subject to rules for physicists. When necessary, he lent me the key or walked down the stair to unlock the heavy doors that led out onto Free School Lane.

Hugh was not in the lab when late on a midsummer June night I went back to shut down the X-ray tube and to develop the photograph of a new TMV sample. It was tilted at about 25 degrees, so that if I were lucky I'd find the helical reflections. The moment I held the still-wet negative against the light box, I knew we had it. The telltale helical markings were unmistakable. Now there should be no problem in persuading Luria and Delbrück that my staying in Cambridge made sense. Despite the midnight hour, I had no desire to go back to my room on Tennis Court Road, and happily I walked along the backs for over an hour.

The following morning I anxiously awaited Francis' arrival to confirm the helical diagnosis. When he needed less than ten seconds to spot the crucial reflection, all my lingering doubts vanished. In fun I went on to trap Francis into believing that I did not think my X-ray picture was in fact very critical. Instead, I argued that the really important step was the cozy-corner insight. These flippant words were hardly out of my mouth before Francis was off on the dangers of uncritical teleology. Francis always said what he meant and assumed that I acted the same way. Though success in Cambridge conversation frequently came from saying something preposterous, hoping that someone would take you seriously, there was no need for Francis to adopt this gambit. A discourse of only one or two minutes on the emotional problems of foreign girls was always sufficient tonic for even the most staid Cambridge evening.

It was, of course, clear what we should next conquer. No more dividends could come quickly from TMV. Further unraveling of its detailed structure needed a more professional attack than I could muster. Moreover, it was not obvious that even the most backbreaking effort would give within several years the structure of the RNA component. The way to DNA was not through TMV.

The moment was thus appropriate to think seriously about some curious regularities in DNA chemistry first observed at Columbia by the Austrian-born biochemist Erwin Chargaff. Since the war, Chargaff and his students had been painstakingly analyzing various DNA samples for the relative proportions of their purine and pyrimidine bases. In all their DNA preparations the number of adenine (A) molecules was very similar to the number of thymine (T) molecules, while the number of guanine (G) molecules was very close to the number of cytosine (C) molecules. Moreover, the proportion of adenine and thymine groups varied with their biological origin. The DNA of some organisms had an excess of A and T, while in other forms of life there was an excess of G and C. No explanation for his striking results was offered by Chargaff, though he obviously thought they were significant. When I first reported them to Francis they did not ring a bell, and he went on thinking about other matters.

Soon afterwards, however, the suspicion that the regularities were important clicked inside his head as the result of several conversations with the young theoretical chemist John Griffith. One occurred while they were drinking beer after an evening talk by the astronomer Tommy Gold on "the perfect cosmological principle."

Tommy's facility for making a far-out idea seem plausible set Francis to wondering whether an argument could be made for a "perfect biological principle." Knowing that Griffith was interested in theoretical schemes for gene replication, he popped out with the idea that the perfect biological principle was the self-replication of the gene—that is, the ability of a gene to be exactly copied when the chromosome number doubles during cell division. Griffith, however, did not go along, since for some months he had preferred a scheme where gene copying was based upon the alternative formation of complementary surfaces.

This was not an original hypothesis. It had been floating about for almost thirty years in the circle of theoretically inclined geneticists intrigued by gene duplication. The argument went that gene duplication required the formation of a complementary (negative) image where shape was related to the original (positive) surface like a lock to a key. The complementary negative image would then function as the mold (template) for the synthesis of a new positive image. A smaller number of geneticists, however, balked at complementary replication. Prominent among them was H. J. Muller, who was impressed that several well-known theoretical physicists, especially Pascual Jordan, thought forces existed by which like attracted like. But Pauling abhorred this direct mechanism and was especially irritated by the suggestion that it was supported by quantum mechanics. Just before the war, he asked Delbrück (who had drawn his attention to Jordan's papers) to coauthor a note to *Science* firmly stating that quantum mechanics favored a gene-duplicating mechanism involving the synthesis of complementary replicas.

Neither Francis nor Griffith was long satisfied that evening by restatements of well-worn hypotheses. Both knew that the important task was now to pinpoint the attractive forces. Here Francis forcefully argued that specific hydrogen bonds were not the answer. They could not provide the necessary exact specificity, since our chemist friends repeatedly told us that the hydrogen atoms in the purine and pyrimidine bases did not have fixed locations but randomly moved from one spot to another. Instead, Francis had the feeling that DNA replication involved specific attractive forces between the flat surfaces of the bases.

Luckily, this was the sort of force that Griffith might just be able to calculate. If the complementary scheme was right, he might find attractive forces between bases with different structures. On the other hand, if direct copying existed, his calculations might reveal attraction between identical bases. Thus, at the closing hour they parted with the understanding that Griffith would see if the calculations were feasible. Several days later, when they bumped into each other in the Cavendish tea queue, Francis learned that a semirigorous argument hinted that adenine and thymine should stick to each other by their flat surfaces. A similar argument could be put forward for attractive forces between guanine and cytosine.

Francis immediately jumped at the answer. If his memory served him right, these were the pairs of bases that Chargaff had shown to occur in equal amounts. Excitedly he told Griffith that I had recently muttered to him some odd results of Chargaff's. At the moment, though, he wasn't sure that the same base pairs were involved. But as soon as the data were checked he would drop by Griffith's rooms to set him straight.

At lunch I confirmed that Francis had got Chargaff's results right. But by then he was only routinely enthusiastic as he went over Griffith's quantum-mechanical arguments. For one thing, Griffith, when pressed, did not want to defend his exact reasoning too strongly. Too many variables had been ignored to make the calculations possible in a reasonable time. Moreover, though each base has two flat sides, no explanation existed for why only one side would be chosen. And there was no reason for ruling out the idea that Chargaff's regularities had their origin in the genetic code. In some way specific groups of nucleotides must code for specific amino acids. Conceivably, adenine equaled thymine because of a yet undiscovered role in the ordering of the bases. There was in addition Roy Markham's assurance that, if Chargaff said that guanine equaled cytosine, he was equally certain it did not. In Markham's eyes, Chargaff's experimental methods inevitably underestimated the true amount of cytosine.

Nonetheless, Francis was not yet ready to dump Griffith's scheme when, early in July, John Kendrew walked into our newly acquired office to tell us that Chargaff himself would soon be in Cambridge for an evening. John had arranged for him to have dinner at Peterhouse, and Francis and I were invited to join them later for drinks in John's room. At High Table John kept the conversation away from serious matters, letting loose only the possibility that Francis and I were going to solve the DNA structure by model building. Chargaff, as one of the world's experts on DNA, was at first not amused by dark horses trying to win the race. Only when John reassured him by mentioning that I was not a typical American did he realize that he was about to listen to a nut. Seeing me quickly reinforced his intuition. Immediately he derided my hair and accent, for since I came from Chicago I had no right to act otherwise. Blandly telling him that I kept my hair long to avoid confusion with American Air Force personnel proved my mental instability.

The high point in Chargaff's scorn came when he led Francis into admitting that he did not remember the chemical differences among the four bases. The faux pas slipped out when Francis mentioned Griffith's calculations. Not remembering which of the bases had amino groups, he could not qualitatively describe the quantum-mechanical argument until he asked Chargaff to write out their formulas. Francis' subsequent retort that he could always look them up got nowhere in persuading Chargaff that we knew where we were going or how to get there.

But regardless of what went through Chargaff's sarcastic mind, someone had to explain his results. Thus the next afternoon Francis buzzed over to Griffith's rooms in Trinity to set himself straight about the base-pair data. Hearing "Come in," he opened the door to see Griffith and a girl. Realizing that this was not the moment for science, he slowly retreated, asking Griffith to tell him again the pairs produced by his calculations. After scribbling them down on the back of an envelope, he left. Since I had departed that morning for the continent, his next stop was the Philosophical Library, where he could remove his lingering doubts about Chargaff's data. Then with both sets of information firmly in hand, he considered returning the next day to Griffith's rooms. But on second thought he realized that Griffith's interests were elsewhere. It was all too clear that the presence of popsies does not inevitably lead to a scientific future.

19

Two weeks later Chargaff and I glanced at each other in Paris. Both of us were there for the International Biochemical Congress. A trace of a sardonic smile was all the recognition I got when we passed in the courtyard outside the massive Salle Richelieu of the Sorbonne. That day I was tracking down Max Delbrück. Before I had left Copenhagen for Cambridge, he had offered me a research position in the biology division of Cal Tech and arranged a Polio Foundation fellowship, to start in September 1952. This March, however, I had written Delbrück that I wanted another Cambridge year. Without any hesitation he saw to it that my forthcoming fellowship was transferred to the Cavendish. Delbrück's speedy approval pleased me, for he had ambivalent feelings about the ultimate value to biology of Pauling-like structural studies.

With the helical TMV picture now in my pocket, I felt more confident that Delbrück would at last wholeheartedly approve my liking for Cambridge. A few minutes' conversation, nonetheless, revealed no basic change in his outlook. Almost no comments emerged from Delbrück as I outlined how TMV was put together. The same indifferent response accompanied my hurriedly delivered summary of our attempts to get DNA by model building. Delbrück was drawn out only by my remark that Francis was exceedingly bright. Unfortunately, I went on to liken Francis' way of thinking to Pauling's. But in Delbrück's world no chemical thought matched the power of a genetic cross. Later that evening, when the geneticist Boris Ephrussi brought up my love affair with Cambridge, Delbrück threw up his hands in disgust.

The sensation of the meeting was the unexpected appearance of Linus. Possibly because there had been considerable newspaper play on the withdrawal of his passport, the State Department reversed itself and allowed Linus to show off the α-helix. A lecture was hastily arranged for the session at which Perutz spoke. Despite the short notice, an overflow crowd was on hand, hoping that they would be the first to learn of a new inspiration. Pauling's talk, however, was only a humorous rehash of published ideas. It nonetheless satisfied everybody, except the few of us who knew his recent papers backward and forward. No new fireworks went off, nor was there any indication given about what now occupied his mind. After his lecture, swarms of admirers surrounded him, and I didn't have the courage to break in before he and his wife, Ava Helen, went back to the nearby Trianon Hotel.

Maurice was about, looking somewhat sour. He had stopped over on his way to Brazil, where he was to lecture for a month on biophysics. His presence surprised me, since it was against his character to seek the trauma of watching two thousand bread-and-butter biochemists pile in and out of badly lighted baroque lecture halls. Speaking down to the cobblestones, he asked me whether I found the talks as tedious as he did. A few academics like Jacques Monod and Sol Spiegelman were enthusiastic speakers, but generally there was so much droning that he found it hard to stay alert for the new facts he should pick up.

I tried to rescue Maurice's morale by bringing him out to the Abbaye at Royaumont for the week-long meeting on phage following the biochemical congress. Though his departure for Rio would limit him to only a night's stay, he liked the idea of meeting the people who did clever biological experiments about DNA. In the train going to Royaumont, however, he looked off-color, giving no indication of wanting either to read *The Times* or to hear me gossip about the phage group. After we were fixed with beds in the high-ceilinged rooms of the partially restored Cistercian monastery, I began talking with some friends I had not seen since leaving the States. Later I kept expecting Maurice to search me out, and when he missed dinner I went up to his room. There I found him lying flat on his stomach, hiding his face from the dim light I had turned on. Something eaten in Paris had not gone down properly, but he told me not to be bothered. The following morning I was given a note saying that he had recovered but had to catch the early train to Paris and apologizing for the trouble he had given me.

Later that morning Lwoff mentioned that Pauling was coming out for a few hours the next day. Immediately I began to think of ways that would allow me to sit next to him at lunch. His visit, however, bore no relation to science. Jeffries Wyman, our scientific attaché in Paris and an acquaintance of Pauling's, thought that Linus and Ava Helen would enjoy the austere charm of the thirteenth-century buildings. During a break in the morning session I caught sight of Wyman's bony, aristocratic face in search of André Lwoff. The Paulings were here and soon began talking to the Delbrücks. Briefly I had Linus to myself after Delbrück mentioned that twelve months hence I was coming to Cal Tech. Our conversation centered on the possibility that at Pasadena I might continue X-ray work with viruses. Virtually no words went to DNA. When I brought up the X-ray pictures at King's, Linus gave the opinion that very accurate X-ray work of the type done by his associates on amino acids was vital to our eventual understanding of the nucleic acids.

I got much further with Ava Helen. Learning that I would be in Cambridge next year, she talked about her son Peter. Already I knew that Peter was accepted by Bragg to work toward a Ph.D. with John Kendrew. This was despite the fact that his Cal Tech grades left much to be desired, even considering his long bout with mononucleosis. John, however, did not want to challenge Linus' desire to place Peter with him, especially knowing that he and his beautiful blonde sister gave smashing parties. Peter and Linda, if she were to visit him, would undoubtedly liven up the Cambridge scene. Then the dream of virtually every Cal Tech chemistry student was that Linda would make his reputation by marrying him. The scuttlebutt about Peter centered on girls and was confused. But now Ava Helen gave me the dope that Peter was an exceptionally fine boy whom everybody would enjoy having around as much as she did. All the same, I remained silently unconvinced that Peter would add as much to our lab as Linda. When Linus beckoned that they must go, I told Ava Helen that I would help her son adjust to the restricted life of the Cambridge research student.

A garden party at Sans Souci, the country home of the Baroness Edmond de Rothschild, effectively brought the meeting to its end. Dressing was no easy matter for me.

Just before the biochemical congress all my belongings were snatched from my train compartment as I was sleeping. Except for a few items picked up at an army PX, the clothes I still possessed had been chosen for a subsequent visit to the Italian Alps. While I felt at ease giving my talk on TMV in shorts, the French contingent feared that I would go one step further by arriving at Sans Souci in the same outfit. A borrowed jacket and tie, however, made me superficially presentable as our bus driver let us out in front of the huge country house.

Sol Spiegelman and I went straight for a butler carrying smoked salmon and champagne, and after a few minutes sensed the value of a cultivated aristocracy. Just before we were to reboard the bus, I wandered into the large drawing room dominated by a Hals and a Rubens. The Baroness was telling several visitors how pleased she was to have such distinguished guests. She did regret, however, that the mad Englishman from Cambridge had decided not to come and enliven the mood. For an instant I was puzzled, until I realized that Lwoff had thought it prudent to warn the Baroness about an unclothed guest who might prove eccentric. The message of my first meeting with the aristocracy was clear. I would not be invited back if I acted like everyone else.

20

To Francis' dismay, I showed little tendency to concentrate on DNA when my summer holiday ended. I was preoccupied with sex, but not of a type that needed encouragement. The mating habits of bacteria were admittedly a unique conversation piece—absolutely no one in his and Odile's social circle would guess bacteria had sex lives. On the other hand, working out how they did it was best left to minor minds. Rumors of male and female bacteria were floating about at Royaumont, but not until early in September, when I attended a small meeting on microbial genetics at Pallanza, did I get the facts from the horse's mouth. There, Cavalli-Sforza and Bill Hayes talked about the experiments by which they and Joshua Lederberg had just established the existence of two discrete bacterial sexes.

Bill's appearance was the sleeper of the three-day gathering: before his talk no one except Cavalli-Sforza knew he existed. As soon as he had finished his unassuming report, however, everyone in the audience knew that a bombshell had exploded in the world of Joshua Lederberg. In 1946 Joshua, then only twenty, burst upon the biological world by announcing that bacteria mated and showed genetic recombination. Since then he had carried out such a prodigious number of pretty experiments that virtually no one except Cavalli dared to work in the same field. Hearing Joshua give Rabelaisian nonstop talks of three to five hours made it all too clear that he was an *enfant terrible*. Moreover, there was his godlike quality of each year expanding in size, perhaps eventually to fill the universe.

Despite Joshua's fabulous cranium, the genetics of bacteria became messier each year. Only Joshua took any enjoyment from the rabbinical complexity shrouding his recent papers. Occasionally I would try to plow through one, but inevitably I'd get stuck and put it aside for another day. No high-power thoughts, however, were required to understand that the discovery of the two sexes might soon make the genetic analysis of bacteria straightforward. Conversations with Cavalli, nonetheless, hinted that Joshua was not yet prepared to think simply. He liked the classical genetic assumption that male and female cells contributed equal amounts of genetic material, even though the resulting analysis was perversely complex. In contrast, Bill's reasoning started from the seemingly arbitrary hypothesis that only a fraction of the male chromosomal material enters the female cell. Given this assumption, further reasoning was infinitely simpler.

As soon as I returned to Cambridge, I beelined out to the library containing the journals to which Joshua had sent his recent work. To my delight I made sense of almost all the previously bewildering genetic crosses. A few matings still were inexplicable, but, even so, the vast masses of data now falling into place made me certain that we were on the right track. Particularly pleasing was the possibility that Joshua might be so stuck on his classical way of thinking that I would accomplish the unbelievable feat of beating him to the correct interpretation of his own experiments.

My desire to clean up skeletons in Joshua's closet left Francis almost cold. The discovery that bacteria were divided into male and female sexes amused but did not arouse him. Almost all his summer had been spent collecting pedantic data for his thesis, and now he was in a mood to think about important facts. Frivolously worrying whether bacteria had one, two, or three chromosomes would not help us win the DNA structure. As long as I kept watch on the DNA literature, there was a chance that something might pop out of lunch or teatime conversations. But if I went back to pure biology, the advantage of our small headstart over Linus might suddenly vanish.

At this time there was still a nagging feeling in Francis' mind that Chargaff's rules were a real key. In fact, when I was away in the Alps he had spent a week trying to prove experimentally that in water solutions there were attractive forces between adenine and thymine, and between guanine and cytosine. But his efforts had yielded nothing. In addition, he was really never at ease talking with Griffith. Somehow their brains didn't jibe well, and there would be long awkward pauses after Francis had thrashed through the merits of a given hypothesis. This was no reason, however, not to tell Maurice that conceivably adenine was attracted to thymine and guanine to cytosine. Since he had to be in London late in October for another reason, he dropped a line to Maurice saying he could come by King's. The reply, inviting him to lunch, was unexpectedly cheerful, and so Francis looked forward to a realistic discussion of DNA.

However, he made the mistake of tactfully appearing not too interested in DNA by starting to talk about proteins. Over half the lunch was thus wasted when Maurice changed the topic to Rosy and droned on and on about her lack of cooperation. Meantime, Francis' mind fastened on a more amusing topic until, the meal over he remembered that he had to rush to a 2:30 appointment. Hurriedly he left the building and was out on the street before realizing he had not brought up the agreement between Griffith's calculations and Chargaff's data. Since it would look to silly to rush back in, he went on, returning that evening to Cambridge. The following morning, after I was told about the futility of the lunch, Francis tried to generate enthusiasm for our having a second go at the structure.

Another zeroing in on DNA, however, did not make sense to me. No fresh facts had come in to chase away the stale taste of last winter's debacle. The only new result we were likely to pick up before Christmas was the divalent metal content of the DNA-containing phage T4. A high value, if found, would strongly suggest binding of Mg^{++} to DNA. With such evidence I might at last force the King's groups to analyze their DNA samples. But the prospects for immediate hard results were not good. First, Maaløe's colleague Nils Jerne must send the phage from Copenhagen. Then I would need to arrange for accurate measurement of both the divalent metals and the DNA content. Finally, Rosy would have to budge.

Fortunately, Linus did not look like an immediate threat on the DNA front. Peter Pauling arrived with the inside news that his father was preoccupied with schemes for the supercoiling of α-helices in the hair protein, keratin. This was not especially good news to Francis. For almost a year he had been in and out of euphoric moods about how α-helices packed together in coiled coils. The trouble was that his mathematics never gelled tightly. When pressed he admitted that his argument had a woolly component. Now he faced the possibility that Linus' solution would be no better and yet he would get all the credit for the coiled coils.

Experimental work for his thesis was broken off so that the coiled-coil equations could be taken up with redoubled effort. This time the correct equations fell out, partly thanks to the help of Kreisel, who had come over to Cambridge to spend a weekend with Francis. A letter to *Nature* was quickly drafted and given to Bragg to send on to the editors, with a covering note asking for speedy publication. If the editors were told that a British article was of above-average interest, they would try to publish the manuscript almost immediately. With luck, Francis' coiled coils would get into print as soon as if not before Pauling's.

Thus there was growing acceptance both in and outside Cambridge that Francis' brain was a genuine asset. Though a few dissidents still thought he was a laughing talking-machine, he nonetheless saw problems through to the finish line. A reflection of his increasing stature was an offer received early in the fall to join David Harker in Brooklyn for a year. Harker, having collected a million dollars to solve the structure of the enzyme ribonuclease, was in search of talent, and the offer of six thousand for one year seemed to Odile wonderfully generous. As expected, Francis had mixed feelings. There must be reasons why there were so many jokes about Brooklyn. On the other hand, he had never been in the States, and even Brooklyn would provide a base from which he could visit more agreeable regions. Also, if Bragg knew that Crick would be away for a year, he might view more favorably a request from Max and John that Francis be reappointed for another three years after his thesis was submitted. The best course seemed tentatively to accept the offer, and in mid-October he wrote Harker that he would come to Brooklyn in the fall of the following year.

As the fall progressed, I remained ensnared by bacterial matings, often going up to London to talk with Bill Hayes at his Hammersmith Hospital lab. My mind snapped back to DNA on the evenings when I managed to catch Maurice for dinner on my way home to Cambridge. Some afternoons, however, he would quietly slip away, and his lab group had it that a special girl friend existed. Finally it came out that everything was above board. The afternoons were spent at a gymnasium learning how to fence.

The situation with Rosy remained as sticky as ever. Upon his return from Brazil, the unmistakable impression was given that she considered collaboration even more impossible than before. Thus, for relief, Maurice had taken up interference microscopy to find a trick for weighing chromosomes. The question of finding Rosy a job elsewhere had been brought to his boss, Randall, but the best to be hoped for would be a new position starting a year hence. Sacking her immediately on the basis of her acid smile just couldn't be done. Moreover, her X-ray pictures were getting prettier and prettier. She gave no sign, however, of liking helices any better. In addition, she thought there was evidence that the sugar-phosphate backbone was on the outside of the molecule. There was no easy way to judge whether this assertion had any scientific basis. As long as Francis and I remained closed out from the experimental data, the best course was to maintain an open mind. So I returned to my thoughts about sex.

21

I WAS by now living in Clare College. Soon after my arrival at the Cavendish, Max had slipped me into Clare as a research student. Working for another Ph.D. was nonsense, but only by using this dodge would I have the possibility of college rooms. Clare was an unexpectedly happy choice. Not only was it on the Cam with a perfect garden but, as I was to learn later, it was especially considerate toward Americans.

Before this happened I was almost stuck in Jesus. At short notice Max and John thought I would have the best chance to be accepted by one of the small colleges, since they had relatively fewer research students than the large, more prestigious, and wealthy colleges like Trinity or King's. Max thus asked the physicist Denis Wilkinson, then a Fellow of Jesus, whether an opening might exist in his college. The following day Denis came by to say Jesus would have me and that I should arrange an appointment to learn the formalities of matriculation.

A talk with its head tutor, however, made me try elsewhere. Jesus' possession of only a few research students appeared related to its formidable reputation on the river. No research student could live in, and so the only predictable consequences of being a Jesus man were bills for a Ph.D. that I would never acquire. Nick Hammond, the classicist head tutor at Clare, painted a much rosier outlook for their foreign research students. In my second year up, I could move into the college. Moreover, at Clare there would be several American research students I might meet.

Nonetheless, during my first Cambridge year, when I lived on Tennis Court Road with the Kendrews, I saw virtually nothing of college life. After matriculation I went into hall for several meals until I discovered that I was unlikely to meet anyone during the ten-to-twelve-minute interval needed to slop down the brown soup, stringy meat, and heavy pudding provided on most evenings. Even during my second Cambridge year, when I moved into rooms on the R staircase of Clare's Memorial Court, my boycott of college food continued. Breakfast at the Whim could occur much later than if I went to hall. For 3/6 the Whim gave a half-warm site to read *The Times* while flat-capped Trinity types turned the pages of the *Telegraph* or *News Chronicle*. Finding suitable evening food on the town was trickier. Eating at the Arts or the Bath Hotel was reserved for special occasions, so when Odile or Elizabeth Kendrew did not invite me to supper I took in the poison put out by the local Indian and Cypriote establishments.

My stomach lasted only until early November before violent pains hit me almost every evening. Alternative treatments with baking soda and milk did not help, and so, despite Elizabeth's assurance that nothing was wrong, I showed up at the ice-cold Trinity Street surgery of a local doctor. After I was allowed to appreciate the oars on his walls, I was expelled with a prescription for a large bottle of white fluid to be taken after meals. This kept me going for almost two weeks, when, with the bottle empty, I returned to the surgery with the fear that I had an ulcer. The news that an alien's dyspeptic pains were persisting did not, however, evoke any sympathetic words, and again I retreated into Trinity Street with a prescription for more white stiff.

That evening I stopped by at the Cricks' newly bought house, hoping that gossip with Odile would make me forget my stomach. The Green Door recently had been abandoned for larger quarters on nearby Portugal Place. Already the dreary wallpaper on the lower floors was gone, and Odile was busy making curtains appropriate for a house large enough to have a bathroom. After I was given a glass of warm milk we began discussing Peter Pauling's discovery of Nina, Max's young Danish *au pair* girl. Then the problem was taken up of how I might establish a connection with the high-class boarding house run by Camille "Pop" Prior at 8 Scroope Terrace. The food at Pop's would offer no improvement over hall, but the French girls who came to Cambridge to improve their English were another matter. A seat at Pop's dinner table, however, could not be asked for directly. Instead, both Odile and Francis thought the best tactic for getting a foot in the door was to commence French lessons with Pop, whose deceased husband had been the Professor of French before the war. If I suited Pop's fancy, I might be invited to one of her sherry parties and meet her current crop of foreign girls. Odile promised to ring Pop to see if lessons could be arranged, and I cycled back to college with the hope that soon my stomach pains would have reason to vanish.

Back in my rooms I lit the coal fire, knowing there was no chance that the sight of my breath would disappear before I was ready for bed. With my fingers too cold to write legibly I huddled next to the fireplace, daydreaming about how several DNA chains could fold together in a pretty and hopefully scientific way. Soon, however, I abandoned thinking at the molecular level and turned to the much easier job of reading biochemical papers on the interrelations of DNA, RNA, and protein synthesis.

Virtually all the evidence then available made me believe that DNA was the template upon which RNA chains were made. In turn, RNA chains were the likely candidates for the templates for protein synthesis. There were some fuzzy data using sea urchins, interpreted as a transformation of DNA into RNA, but I preferred to trust other experiments showing that DNA molecules, once synthesized, are very very stable. The idea of the genes' being immortal smelled right, and so on the wall above my desk I taped up a paper sheet saying DNA → RNA → protein. The arrows did not signify chemical transformations, but instead expressed the transfer of genetic information from the sequences of nucleotides in DNA molecules to the sequences of amino acids in proteins.

Though I fell asleep contented with the thought that I understood the relationship between nucleic acids and protein synthesis, the chill of dressing in an ice-cold bedroom brought me back to the knowing truth that a slogan was no substitute for the DNA structure. Without it, the only impact that Francis and I were likely to have was to convince the biochemists we met in a nearby pub that we would never appreciate the fundamental significance of complexity in biology. What was worse, even when Francis stopped thinking about coiled coils or I about bacterial genetics, we still remained stuck at the same place we were twelve months before. Lunches at the Eagle frequently went by without a mention of DNA, though usually somewhere on our after-lunch walk along the backs genes would creep in for a moment.

On a few walks our enthusiasm would build up to the point that we fiddled with the models when we got back to our office. But almost immediately Francis saw that the reasoning which had momentarily given us hope led nowhere. Then he would go back to the examination of the hemoglobin X-ray photographs out of which his thesis must emerge. Several times I carried on alone for a half hour or so, but without Francis' reassuring chatter my inability to think in three dimensions became all too apparent.

I was thus not at all displeased that we were sharing our office with Peter Pauling, then living in the Peterhouse hostel as a research student of John Kendrew's. Peter's presence meant that, whenever more science was pointless, the conversation could dwell on the comparative virtues of girls from England, the Continent, and California. A fetching face, however, had nothing to do with the broad grin on Peter's face when he sauntered into the office one afternoon in the middle of December and put his feet up on his desk. In his hand was a letter from the States that he had picked up on his return to Peterhouse for lunch.

It was from his father. In addition to routine family gossip was the long-feared news that Linus now had a structure for DNA. No details were given of what he was up to, and so each time the letter passed between Francis and me the greater was our frustration. Francis then began pacing up and down the room thinking aloud, hoping that in a great intellectual fervor he could reconstruct what Linus might have done. As long as Linus had not told us the answer, we should get equal credit if we announced it at the same time.

Nothing worthwhile had emerged, though, by the time we walked upstairs to tea and told Max and John of the letter. Bragg was in for a moment, but neither of us wanted the perverse joy of informing him that the English labs were again about to be humiliated by the Americans.

As we munched chocolate biscuits, John tried to cheer us up with the possibility of Linus' being wrong. After all, he had never seen Maurice's and Rosy's pictures. Our hearts, however, told us otherwise.

22

NO FURTHER news emerged from Pasadena before Christmas. Our spirits slowly went up, for if Pauling had found a really exciting answer the secret could not be kept long. One of his graduate students must certainly know what his model looked like, and if there were obvious biological implications the rumor would have quickly reached us. Even if Linus was somewhere near the right structure, the odds seemed against his getting near the secret of gene replication. Also, the more we thought about DNA chemistry, the more unlikely seemed the possibility that even Linus could pick off the structure in total ignorance of the work at King's.

Maurice was told that Pauling was in his pasture when I passed through London on my way to Switzerland for a Christmas skiing holiday. I was hoping that the urgency created by Linus' assault on DNA might make him ask Francis and me for help. However, if Maurice thought that Linus had a chance to steal the prize, he didn't let on. Much more important was the news that Rosy's days at King's were numbered. She had told Maurice that she wanted soon to transfer to Bernal's lab at Birkbeck College. Moreover, to Maurice's surprise and relief, she would not take the DNA problem with her. In the next several months she was to conclude her stay by writing up her work for publication. Then, with Rosy at last out of his life, he would commence an all-out search for the structure.

Upon my return to Cambridge in mid-January, I sought out Peter to learn what was in his recent letters from home. Except for one brief reference to DNA, all the news was family gossip. The one pertinent item, however, was not reassuring. A manuscript on DNA had been written, a copy of which would soon be sent to Peter. Again there was not a hint of what the model looked like. While waiting for the manuscript to arrive, I kept my nerves in check by writing up my ideas on bacterial sexuality. A quick visit to Cavalli in Milan, which occurred just after my skiing holiday in Zermatt, had convinced me that my speculations about how bacteria mated were likely to be right. Since I was afraid that Lederberg might soon see the same light, I was anxious to publish quickly a joint article with Bill Hayes. But this manuscript was not in final form when, in the first week of February, the Pauling paper crossed the Atlantic.

Two copies, in fact, were dispatched to Cambridge—one to Sir Lawrence, the other to Peter. Bragg's response upon receiving it was to put it aside. Not knowing that Peter would also get a copy, he hesitated to take the manuscript down to Max's office. There Francis would see it and set off on another wild-goose chase. Under the present timetable there were only eight months more of Francis' laugh to bear. That is, if his thesis was finished on schedule. Then for a year, if not more, with Crick in exile in Brooklyn, peace and serenity would prevail.

While Sir Lawrence was pondering whether to chance taking Crick's mind off his thesis, Francis and I were poring over the copy that Peter brought in after lunch. Peter's face betrayed something important as he entered the door, and my stomach sank in apprehension at learning that all was lost. Seeing that neither Francis nor I could bear any further suspense, he quickly told us that the model was a three-chain helix with the sugar-phosphate backbone in the center. This sounded so suspiciously like our aborted effort of last year that immediately I wondered whether we might already have had the credit and glory of a great discovery if Bragg had not held us back. Giving Francis no chance to ask for the manuscript, I pulled it out of Peter's outside coat pocket and began reading. By spending less than a minute with the summary and the introduction, I was soon at the figures showing the locations of the essential atoms.

At once I felt something was not right. I could not pinpoint the mistake, however, until I looked at the illustrations for several minutes. Then I realized that the phosphate groups in Linus' model were not ionized, but that each group contained a bound hydrogen atom and so had no net charge. Pauling's nucleic acid in a sense was not an acid at all. Moreover, the uncharged phosphate groups were not incidental features. The hydrogens were part of the hydrogen bonds that held together the three intertwined chains. Without the hydrogen atoms, the chains would immediately fly apart and the structure vanish.

Everything I knew about nucleic-acid chemistry indicated that phosphate groups never contained bound hydrogen atoms. No one had ever questioned that DNA was a moderately strong acid. Thus, under physiological conditions, there would always be positively charged ions like sodium or magnesium lying nearby to neutralize the negatively charged phosphate groups. All our speculations about whether divalent ions held the chains together would have made no sense if there were hydrogen atoms firmly bound to the phosphates. Yet somehow Linus, unquestionably the world's most astute chemist, had come to the opposite conclusion.

When Francis was amazed equally by Pauling's unorthodox chemistry, I began to breathe slower. By then I knew we were still in the game. Neither of us, however, had the slightest clue to the steps that had led Linus to his blunder. If a student had made a similar mistake, he would be thought unfit to benefit from Cal Tech's chemistry faculty. Thus, we could not but initially worry whether Linus' model followed from a revolutionary reevaluation of the acid-base properties of very large molecules. The tone of the manuscript, however, argued against any such advance in chemical theory. No reason existed to keep secret a first-rate theoretical breakthrough. Rather, if that had occurred Linus would have written two papers, the first describing his new theory, the second showing how it was used to solve the DNA structure.

The blooper was too unbelievable to keep secret for more than a few minutes. I dashed over to Roy Markham's lab to spurt out the news and to receive further reassurance that Linus' chemistry was screwy. Markham predictably expressed pleasure that a giant had forgotten elementary college chemistry. He then could not refrain from revealing how one of Cambridge's great men had on occasion also forgotten his chemistry. Next I hopped over to the organic chemists', where again I heard the soothing words that DNA was an acid.

By teatime I was back in the Cavendish, where Francis

was explaining to John and Max that no further time must be lost on this side of the Atlantic. When his mistake became known, Linus would not stop until he had captured the right structure. Now our immediate hope was that his chemical colleagues would be more than ever awed by his intellect and not probe the details of his model. But since the manuscript had already been dispatched to the *Proceedings of the National Academy,* by mid-March at the latest Linus' paper would be spread around the world. Then it would be only a matter of days before the error would be discovered. We had anywhere up to six weeks before Linus again was in full-time pursuit of DNA.

Though Maurice had to be warned, we did not immediately ring him. The pace of Francis' words might cause Maurice to find a reason for terminating the conversation before all the implications of Pauling's folly could be hammered home. Since in several days I was to go up to London to see Bill Hayes, the sensible course was to bring the manuscript with me for Maurice's and Rosy's inspection.

Then, as the stimulation of the last several hours had made further work that day impossible, Francis and I went over to the Eagle. The moment its doors opened for the evening we were there to drink a toast to the Pauling failure. Instead of sherry, I let Francis buy me a whiskey. Though the odds still appeared against us, Linus had not yet won his Nobel.

23

MAURICE was busy when, just before four, I walked in with the news that the Pauling model was far off base. So I went down the corridor to Rosy's lab, hoping she would be about. Since the door was already ajar, I pushed it open to see her bending over a lighted box upon which lay an X-ray photograph she was measuring. Momentarily startled by my entry, she quickly regained her composure and, looking straight at my face, let her eyes tell me that uninvited guests should have the courtesy to knock.

I started to say that Maurice was busy, but before the insult was out I asked her whether she wanted to look at Peter's copy of his father's manuscript. Though I was curious how long she would take to spot the error, Rosy was not about to play games with me. I immediately explained where Linus had gone astray. In doing so, I could not refrain from pointing out the superficial resemblance between Pauling's three-chain helix and the model that Francis and I had shown her fifteen months earlier. The fact that Pauling's deductions about symmetry were no more inspired than our awkward efforts of the year before would, I thought, amuse her. The result was just the opposite. Instead, she became increasingly annoyed with my recurring references to helical structures. Coolly she pointed out that not a shred of evidence permitted Linus, or anyone else, to postulate a helical structure for DNA. Most of my words to her were superfluous, for she knew that Pauling was wrong the moment I mentioned a helix.

Interrupting her harangue, I asserted that the simplest form for any regular polymeric molecule was a helix. Knowing that she might counter with the fact that the sequence of bases was unlikely to be regular, I went on with the argument that, since DNA molecules form crystals, the nucleotide order must not affect the general structure. Rosy by then was hardly able to control her temper, and her voice rose as she told me that the stupidity of my remarks would be obvious if I would stop blubbering and look at her X-ray evidence.

I was more aware of her data than she realized. Several months earlier Maurice had told me the nature of her so-called antihelical results. Since Francis had assured me that they were a red herring, I decided to risk a full explosion. Without further hesitation I implied that she was incompetent in interpreting X-ray pictures. If only she would learn some theory, she would understand how her supposed antihelical features arose from the minor distortions needed to pack regular helices into a crystalline lattice.

Suddenly Rosy came from behind the lab bench that separated us and began moving toward me. Fearing that in her hot anger she might strike me, I grabbed up the Pauling manuscript and hastily retreated to the open door. My escape was blocked by Maurice, who, searching for me, had just then stuck his head through. While Maurice and Rosy looked at each other over my slouching figure, I lamely told Maurice that the conversation between Rosy and me was over and that I had been about to look for him in the tea room. Simultaneously I was inching my body from between them, leaving Maurice face to face with Rosy. Then, when Maurice failed to disengage himself immediately, I feared that out of politeness he would ask Rosy to join us for tea. Rosy, however, removed Maurice from his uncertainty by turning around and firmly shutting the door.

Walking down the passage, I told Maurice how his unexpected appearance might have prevented Rosy from assaulting me. Slowly he assured me that this very well might have happened. Some months earlier she had made a similar lunge toward him. They had almost come to blows following an argument in his room. When he wanted to escape, Rosy had blocked the door and had moved out of the way only at the last moment. But then no third person was on hand.

My encounter with Rosy opened up Maurice to a degree that I had not seen before. Now that I need no longer merely imagine the emotional hell he had faced during the past two years, he could treat me almost as a fellow collaborator rather than as a distant acquaintance with whom close confidences inevitably led to painful misunderstandings. To my surprise, he revealed that with the help of his assistant Wilson he had quietly been duplicating some of Rosy's and Gosling's X-ray work. Thus there need not be a large time gap before Maurice's research efforts were in full swing. Then the even more important cat was let out of the bag: since the middle of the summer Rosy had had evidence for a new three-dimensional form of DNA. It occurred when the DNA molecules were surrounded by a large amount of water. When I asked what the pattern was like, Maurice went into the adjacent room to pick up a print of the new form they called the "B" structure.

The instant I saw the picture my mouth fell open and my pulse began to race. The pattern was unbelievably simpler than those obtained previously ("A" form). Moreover, the black cross of reflections which dominated the picture could arise only from a helical structure. With the A form, the argument for a helix was never straightforward, and considerable ambiguity existed as to exactly which type of helical symmetry was present. With the B form, however, mere inspection of its X-ray picture gave several of the vital helical parameters. Conceivably, after only a few minutes' calculations, the number of chains in the molecule could be fixed. Pressing Maurice for what they had done using the B photo, I learned that his colleague R. D. B. Fraser earlier had been doing some serious playing with three-chain models but that so far nothing exciting had come up. Though Maurice conceded that the evidence for a helix was now overwhelming—the Stokes-Cochran-Crick theory clearly indicated that a helix must exist—this was not to him of major significance. After all, he had previously thought a helix would emerge. The real problem was the absence of any structural hypothesis which would allow them to pack the bases regularly in the inside of the helix. Of course this presumed that Rosy had hit it right in wanting the bases in the center and the backbone outside. Though Maurice told me he was now quite convinced she was correct, I remained skeptical, for her evidence was still out of the reach of Francis and me.

On our way to Soho for supper I returned to the problem of Linus, emphasizing that smiling too long over his mistake might be fatal. The position would be far safer if Pauling had been merely wrong instead of looking like a fool. Soon, if not already, he would be at it day and night. There was the further danger that if he put one of his assistants to taking DNA photographs, the B structure would also be discovered in Pasadena. Then, in a week at most, Linus would have the structure.

Maurice refused to get excited. My repeated refrain that DNA could fall at any moment sounded too suspiciously like Francis in one of his overwrought periods. For years Francis had been trying to tell him what was important, but the more dispassionately he considered his life, the more he knew he had been wise to follow up his own hunches. As the waiter peered over his shoulder, hoping we would finally order, Maurice made sure I understood that if we could all agree where science was going, everything would be solved and we would have no recourse but to be engineers or doctors.

With the food on the table I tried to fix our thoughts on the chain number, arguing that measuring the location of the innermost reflection on the first and second layer lines might immediately set us on the right track. But since Maurice's long-drawn-out reply never came to the point, I could not decide whether he was saying that no one at King's had measured the pertinent reflections or whether he wanted to eat his meal before it got cold. Reluctantly I ate, hoping that after coffee I might get more details if I walked him back to his flat. Our bottle of Chablis, however, diminished my desire for hard facts, and as we walked out of Soho and across Oxford Street, Maurice spoke only of his plans to get a less gloomy apartment in a quieter area.

Afterwards, in the cold, almost unheated train compartment, I sketched on the blank edge of my newspaper what I remembered of the B pattern. Then as the train jerked toward Cambridge, I tried to decide between two- and three-chain models. As far as I could tell, the reason the King's group did not like two chains was not foolproof. It depended upon the water content of the DNA samples, a value they admitted might be in great error. Thus by the time I had cycled back to college and climbed over the back gate, I had decided to build two-chain models. Francis would have to agree. Even though he was a physicist, he knew that important biological objects come in pairs.

24

BRAGG was in Max's office when I rushed in the next day to blurt out what I had learned. Francis was not yet in, for it was a Saturday morning and he was still home in bed glancing at the *Nature* that had come in the morning mail. Quickly I started to run through the details of the B form, making a rough sketch to show the evidence that DNA was a helix which repeated its pattern every 34 Å along the helical axis. Bragg soon interrupted me with a question, and I knew my argument had got across. I thus wasted no time in bringing up the problem of Linus, giving the opinion that he was far too dangerous to be allowed a second crack at DNA while the people on this side of the Atlantic sat on their hands. After saying that I was going to ask a Cavendish machinist to make models of the purines and pyrimidines, I remained silent, waiting for Bragg's thoughts to congeal.

To my relief, Sir Lawrence not only made no objection but encouraged me to get on with the job of building models. He clearly was not in sympathy with the internal squabbling at King's—especially when it might allow Linus, of all people, to get the thrill of discovering the structure of still another important molecule. Also aiding our cause was my work on tobacco mosaic virus. It had given Bragg the impression that I was on my own. Thus he could fall asleep that night untroubled by the nightmare that he had given Crick carte blanche for another foray into frenzied inconsiderateness. I then dashed down the stairs to the machine shop to warn them that I was about to draw up plans for models wanted within a week.

Shortly after I was back in our office, Francis strolled in to report that their last night's dinner party was a smashing success. Odile was positively enchanted with the French boy that my sister had brought along. A month previously Elizabeth had arrived for an indefinite stay on her way back to the States. Luckily I could both install her in Camille Prior's boarding house and arrange to take my evening meals there with Pop and her foreign girls. Thus in one blow Elizabeth had been saved from typical English digs, while I looked forward to a lessening of my stomach pains.

Also living at Pop's was Bertrand Fourcade, the most beautiful male, if not person, in Cambridge. Bertrand, then visiting for a few months to perfect his English, was not unconscious of his unusual beauty and so welcomed the companionship of a girl whose dress was not in shocking contrast with his well-cut clothes. As soon as I had mentioned that we knew the handsome foreigner, Odile expressed delight. She, like many Cambridge women, could not take her eyes off Bertrand whenever she spotted him walking down King's Parade or standing about looking very well-favored during the intermissions of plays at the amateur dramatic club. Elizabeth was thus given the task of seeing whether Bertrand would be free to join us for a meal with the Cricks at Portugal Place. The time finally arranged, however, had overlapped my visit to London. When I was watching Maurice meticulously finish all the food on his plate, Odile was admiring Bertrand's perfectly proportioned face as he spoke of his problems choosing among potential social engagements during his forthcoming summer on the Riviera.

This morning Francis saw that I did not have my usual interest in the French moneyed gentry. Instead, for a moment he feared that I was going to be unusually tiresome. Reporting that even a former birdwatcher could now solve DNA was not the way to greet a friend bearing a slight hangover. However, as soon as I revealed the B-pattern details, he knew I was not pulling his leg. Especially important was my insistence that the meridional reflection at 3.4 Å was much stronger than any other reflection. This could only mean that the 3.4 Å-thick purine and pyrimidine bases were stacked on top of each other in a direction perpendicular to the helical axis. In addition we could feel sure from both electron-microscope and X-ray evidence that the helix diameter was about 20 Å.

Francis, however, drew the line against accepting my assertion that the repeated finding of twoness in biological systems told us to build two-chain models. The way to get on, in his opinion, was to reject any argument which did not arise from the chemistry of nucleic-acid chains. Since the experimental evidence known to us could not yet distinguish between two- and three-chain models, he wanted to pay equal attention to both alternatives. Though I remained totally skeptical, I saw no reason to contest his words. I would of course start playing with two-chain models.

No serious models were built, however, for several days. Not only did we lack the purine and pyrimidine components, but we had never had the shop put together any phosphorus atoms. Since our machinist needed at least three days merely to turn out the more simple phosphorus atoms, I went back to Clare after lunch to hammer out the final draft of my genetics manuscript. Later, when I cycled over to Pop's for dinner, I found Bertrand and my sister talking to Peter Pauling, who the week before had charmed Pop into giving him dining rights. In contrast to Peter, who was complaining that the Perutzes had no right to keep Nina home on a Saturday night, Bertrand and Elizabeth looked pleased with themselves. They had just returned from motoring in a friend's Rolls to a celebrated country house near Bedford. Their host, an antiquarian architect, had never truckled under to modern civilization and kept his house free of gas and electricity. In all ways possible he maintained the life of an eighteenth-century squire, even to providing special walking sticks for his guests as they accompanied him around his grounds.

Dinner was hardly over before Bertrand whisked Elizabeth on to another party, leaving Peter and me at a loss for something to do. After first deciding to work on his hi-fi set, Peter came along with me to a film. This kept us in check until, as midnight approached, Peter held forth on how Lord Rothschild was avoiding his responsibility as a father by not inviting him to dinner with his daughter Sarah. I could not disagree, for if Peter moved into the fashionable world I might have a chance to escape acquiring a faculty-type wife.

Three days later the phosphorus atoms were ready, and I quickly strung together several short sections of the sugar-phosphate backbone. Then for a day and a half I tried to find a suitable two-chain model with the backbone in the center. All the possible models compatible with the B-form X-ray data, however, looked stereochemically even more unsatisfactory than our three-chained models of fifteen months before. So, seeing Francis absorbed by his thesis, I took off the afternoon to play tennis with Bertrand. After tea I returned to point out that it was lucky I found tennis more pleasing than model building. Francis, totally indifferent to the perfect spring day, immediately put down his pencil to point out that not only was DNA very important, but he could assure me that someday I would discover the unsatisfactory nature of outdoor games.

During dinner at Portugal Place I was back in a mood to worry about what was wrong. Though I kept insisting that we should keep the backbone in the center, I knew none of my reasons held water. Finally over coffee I admitted that my reluctance to place the bases inside partially arose from the suspicion that it would be possible to build an almost infinite number of models of this type. Then we would have the impossible task of deciding whether one was right. But the real stumbling block was the bases. As long as they were outside, we did not have to consider them. If they were pushed inside, the frightful problem existed of how to pack together two or more chains with irregular sequences of bases. Here Francis had to admit that he saw not the slightest ray of light. So when I walked up out of their basement dining room into the street, I left Francis with the impression that he would have to provide at least a semiplausible argument before I would seriously play about with base-centered models.

The next morning, however, as I took apart a particularly repulsive backbone-centered molecule, I decided that no harm could come from spending a few days building backbone-out models. This meant temporarily ignoring the bases, but in any case this had to happen since now another week was required before the shop could hand over the flat tin plates cut in the shapes of purines and pyrimidines.

There was no difficulty in twisting an externally situated backbone into a shape compatible with the X-ray evidence. In fact, both Francis and I had the impression that the most satisfactory angle of rotation between two adjacent bases was between 30 and 40 degrees. In contrast, an angle either twice as large or twice as small looked incompatible with the relevant bond angles. So if the backbone was on the outside, the crystallographic repeat of 34 Å had to represent the distance along the helical axis required for a complete rotation. At this stage Francis' interest began to perk up, and at increasing frequencies he would look up from his calculations to glance at the model. Nonetheless, neither of us had any hesitation in breaking off work for the weekend. There was a party at Trinity on Saturday night, and on Sunday Maurice was coming up to the Cricks' for a social visit arranged weeks before the arrival of the Pauling manuscript.

Maurice, however, was not allowed to forget DNA. Almost as soon as he arrived from the station, Francis started to probe him for fuller details of the B pattern. But by the end of lunch Francis knew no more than I had picked up the week before. Even the presence of Peter, saying he felt sure his father would soon spring into action, failed to ruffle Maurice's plans. Again he emphasized that he wanted to put off more model building until Rosy was gone, six weeks from then. Francis seized the occasion to ask Maurice whether he would mind if we started to play about with DNA models. When Maurice's slow answer emerged as no, he wouldn't mind, my pulse rate returned to normal. For even if the answer had been yes, our model building would have gone ahead.

25

THE next few days saw Francis becoming increasingly agitated by my failure to stick close to the molecular models. It did not matter that before his tenish entrance I was usually in the lab. Almost every afternoon, knowing that I was on the tennis court, he would fretfully twist his head away from his work to see the polynucleotide backbone unattended. Moreover, after tea I would show up for only a few minutes of minor fiddling before dashing away to have sherry with the girls at Pop's. Francis' grumbles did not disturb me, however, because further refining of our latest backbone without a solution to the bases would not represent a real step forward.

I went ahead spending most evenings at the films, vaguely dreaming that any moment the answer would suddenly hit me. Occasionally my wild pursuit of the celluloid backfired, the worst occasion being an evening set aside for *Ecstasy*. Peter and I had both been too young to observe the original showings of Hedy Lamarr's romps in the nude, and so on the long-awaited night we collected Elizabeth and went up to the Rex. However, the only swimming scene left intact by the English censor was an inverted reflection from a pool of water. Before the film was half over we joined the violent booing of the disgusted undergraduates as the dubbed voices uttered words of uncontrolled passion.

Even during good films I found it almost impossible to forget the bases. The fact that we had at last produced a stereochemically reasonable configuration for the backbone was always in the back of my head. Moreover, there was no longer any fear that it would be incompatible with the experimental data. By then it had been checked out with Rosy's precise measurements. Rosy, of course, did not directly give us her data. For that matter, no one at King's realized they were in our hands. We came upon them because of Max's membership on a committee appointed by the Medical Research Council to look into the research activities of Randall's lab. Since Randall wished to convince the outside committee that he had a productive research group, he had instructed his people to draw up a comprehensive summary of their accomplishments. In due time this was prepared in mimeograph form and sent routinely to all the committee members. As soon as Max saw the sections by Rosy and Maurice, he brought the report in to Francis and me. Quickly scanning its contents, Francis sensed with relief that following my return from King's I had correctly reported to him the essential features of the B pattern. Thus only minor modifications were necessary in our backbone configuration.

Generally, it was late in the evening after I got back to my rooms that I tried to puzzle out the mystery of the bases. Their formulas were written out in J. N. Davidson's little book *The Biochemistry of Nucleic Acids,* a copy of which I kept in Clare. So I could be sure that I had the correct structures when I drew tiny pictures of the bases on sheets of Cavendish notepaper. My aim was somehow to arrange the centrally located bases in such a way that the backbones on the outside were completely regular—that is, giving the sugar-phosphate groups of each nucleotide identical three-dimensional configurations. But each time I tried to come up with a solution I ran into the obstacle that the four bases each had a quite different shape. Moreover, there were many reasons to believe that the sequences of the bases of a given polynucleotide chain were very irregular. Thus, unless some very special trick existed, randomly twisting two polynucleotide chains around one another should result in a mess. In some places the bigger bases must touch each other, while in other regions, where the smaller bases would lie opposite each other, there must exist a gap or else their backbone regions must buckle in.

There was also the vexing problem of how the intertwined chains might be held together by hydrogen bonds between the bases. Though for over a year Francis and I had dismissed the possibility that bases formed regular hydrogen bonds, it was now obvious to me that we had done so incorrectly. The observation that one or more hydrogen atoms on each of the bases could move from one location to another (a tautomeric shift) had initially led us to conclude that all the possible tautomeric forms of a given base occurred in equal frequencies. But a recent re-reading of J. M. Gulland's and D. O. Jordan's papers on the acid and base titrations of DNA made me finally appreciate the strength of their conclusion that a large fraction, if not all, of the bases formed hydrogen bonds to other bases. Even more important, these hydrogen bonds were present at very low DNA concentrations, strongly hinting that the bonds linked together bases in the same molecule. There was in addition the X-ray crystallographic result that each pure base so far examined formed as many irregular hydrogen bonds as stereochemically possible. Thus, conceivably the crux of the matter was a rule governing hydrogen bonding between bases.

My doodling of the bases on paper at first got nowhere, regardless of whether or not I had been to a film. Even the necessity to expunge *Ecstasy* from my mind did not lead to passable hydrogen bonds, and I fell asleep hoping that an undergraduate party the next afternoon at Downing would be full of pretty girls. But my expectations were dashed as soon as I arrived to spot a group of healthy hockey players and several pallid debutantes. Bertrand also instantly perceived he was out of place, and as we passed a polite interval before scooting out, I explained how I was racing Peter's father for the Nobel Prize.

Not until the middle of the next week, however, did a nontrivial idea emerge. It came while I was drawing the fused rings of adenine on paper. Suddenly I realized the potentially profound implications of a DNA structure in which the adenine residue formed hydrogen bonds similar to those found in crystals of pure adenine. If DNA was like this, each adenine residue would form two hydrogen bonds to an adenine residue related to it by a 180-degree rotation. Most important, two symmetrical hydrogen bonds could also hold together pairs of guanine, cytosine, or thymine. I thus started wondering whether each DNA molecule consisted of two chains with identical base sequences held together by hydrogen bonds between pairs of identical bases. There was the complication, however, that such a structure could not have a regular backbone, since the purines (adenine and guanine) and the pyrimidines (thymine and cytosine) have different shapes. The

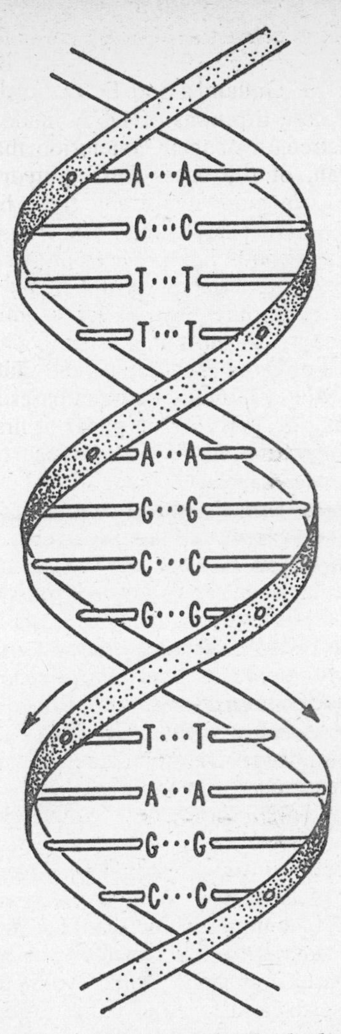

A schematic view of a DNA molecule built up from like-with-like base pairs.

resulting backbone would have to show minor in-and-out buckles depending upon whether pairs of purines or pyrimidines were in the center.

Despite the messy backbone, my pulse began to race. If this was DNA, I should create a bombshell by announcing its discovery. The existence of two intertwined chains with identical base sequences could not be a chance matter. Instead it would strongly suggest that one chain in each molecule had at some earlier stage served as the template for the synthesis of the other chain. Under this scheme, gene replication starts with the separation of its two identical chains. Then two new daughter strands are made on the two parental templates, thereby forming two DNA molecules identical to the original molecule. Thus, the essential trick of gene replication could come from the requirement that each base in the newly synthesized chain always hydrogen-bonds to an identical base. That night, however, I could not see why the common tautomeric form of guanine would not hydrogen-bond to adenine. Likewise, several other pairing mistakes should also occur. But since there was no reason to rule out the participation of specific enzymes, I saw no need to be unduly disturbed. For example, there might exist an enzyme specific for adenine that caused adenine always to be inserted opposite an adenine residue on the template strands.

As the clock went past midnight I was becoming more and more pleased. There had been far too many days when Francis and I worried that the DNA structure might turn out to be superficially very dull, suggesting nothing about either its replication or its function in controlling cell biochemistry. But now, to my delight and amazement, the answer was turning out to be profoundly interesting. For over two hours I happily lay awake with pairs of adenine residues whirling in front of my closed eyes. Only for brief moments did the fear shoot through me that an idea this good could be wrong.

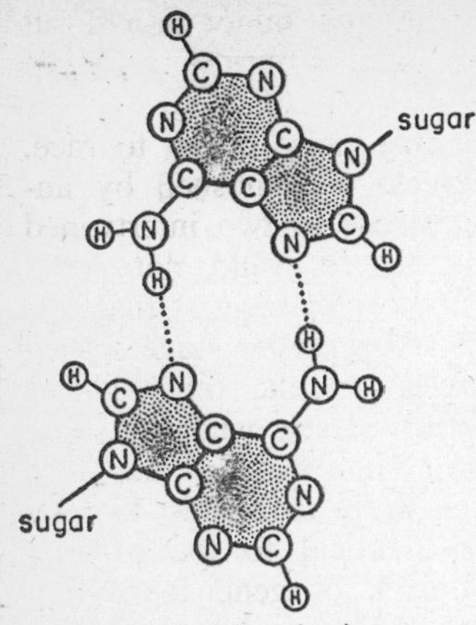

adenine with adenine

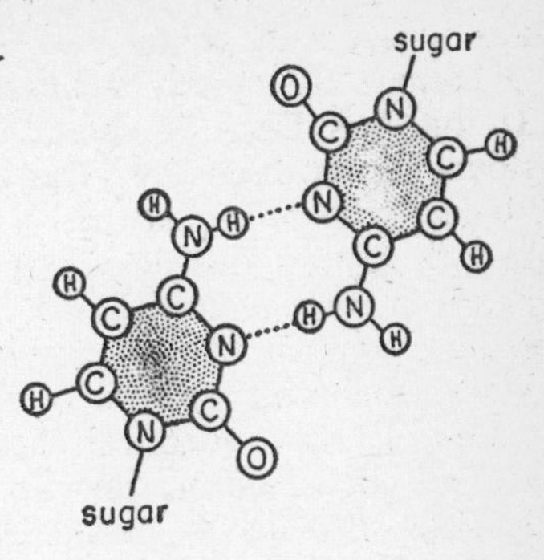

cytosine with cytosine

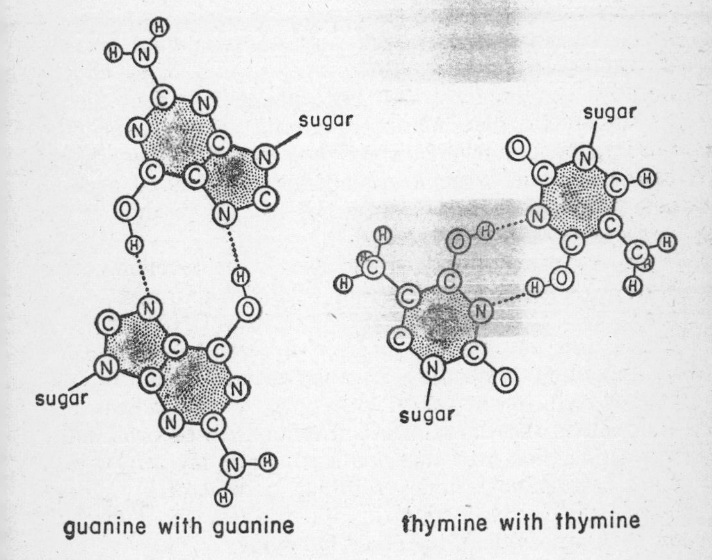

guanine with guanine

thymine with thymine

The four base pairs used to construct the like-with-like structure (hydrogen bonds are dotted).

26

My scheme was torn to shreds by the following noon. Against me was the awkward chemical fact that I had chosen the wrong tautomeric forms of guanine and thymine. Before the disturbing truth came out, I had eaten a hurried breakfast at the Whim, then momentarily gone back to Clare to reply to a letter from Max Delbrück which reported that my manuscript on bacterial genetics looked unsound to the Cal Tech geneticists. Nevertheless, he would accede to my request that he send it to the *Proceedings of the National Academy*. In this way, I would still be young when I committed the folly of publishing a silly idea. Then I could sober up before my career was permanently fixed on a reckless course.

At first this message had its desired unsettling effect. But now, with my spirits soaring on the possibility that I had the self-duplicating structure, I reiterated my faith that I knew what happened when bacteria mated. Moreover, I could not refrain from adding a sentence saying that I had just devised a beautiful DNA structure which was completely different from Pauling's. For a few seconds I considered giving some details of what I was up to, but since I was in a rush I decided not to, quickly dropped the letter in the box, and dashed off to the lab.

The letter was not in the post for more than an hour before I knew that my claim was nonsense. I no sooner got to the office and began explaining my scheme than the American crystallographer Jerry Donohue protested that the idea would not work. The tautomeric forms I had copied out of Davidson's book were, in Jerry's opinion, incorrectly assigned. My immediate retort that several other texts also pictured guanine and thymine in the enol form cut no ice with Jerry. Happily he let out that for years organic chemists had been arbitrarily favoring particular tautomeric forms over their alternatives on only the flimsiest of grounds. In fact, organic-chemistry text-

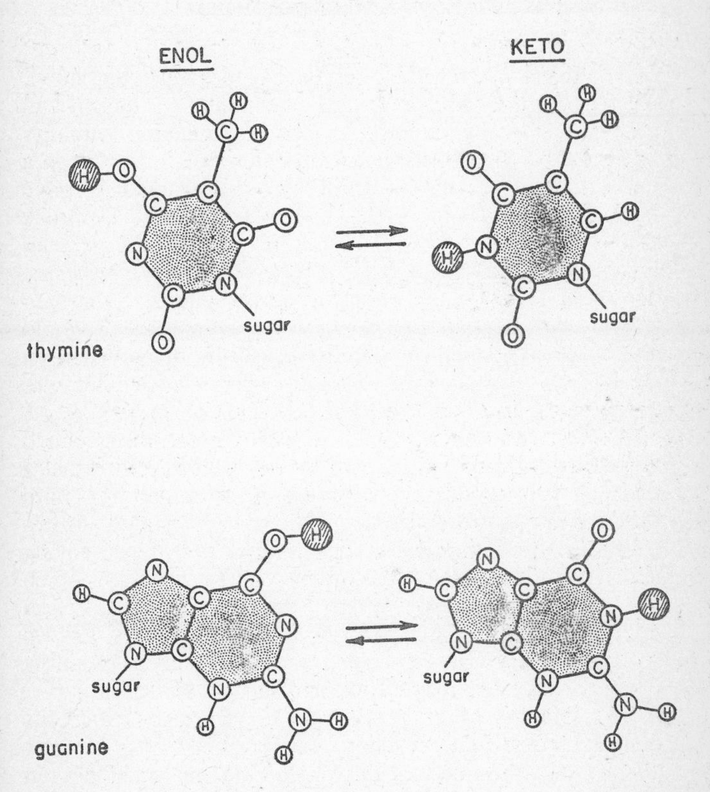

The contrasting tautomeric forms of guanine and thymine which might occur in DNA. The hydrogen atoms that can undergo the changes in position (a tautomeric shift) are shaded.

books were littered with pictures of highly improbable tautomeric forms. The guanine picture I was thrusting toward his face was almost certainly bogus. All his chemical intuition told him that it would occur in the keto form. He was just as sure that thymine was also wrongly assigned an enol configuration. Again he strongly favored the keto alternative.

Jerry, however, did not give a foolproof reason for preferring the keto forms. He admitted that only one crystal structure bore on the problem. This was diketopiperazine, whose three-dimensional configuration had been carefully worked out in Pauling's lab several years before. Here there was no doubt that the keto form, not the enol, was present. Moreover, he felt sure that the quantum-mechanical arguments which showed why diketopiperazine has the keto form should also hold for guanine and thymine. I was thus firmly urged not to waste more time with my harebrained scheme.

Though my immediate reaction was to hope that Jerry was blowing hot air, I did not dismiss his criticism. Next to Linus himself, Jerry knew more about hydrogen bonds than anyone else in the world. Since for many years he had worked at Cal Tech on the crystal structures of small organic molecules, I couldn't kid myself that he did not grasp our problem. During the six months that he occupied a desk in our office, I had never heard him shooting off his mouth on subjects about which he knew nothing.

Thoroughly worried, I went back to my desk hoping that some gimmick might emerge to salvage the like-with-like idea. But it was obvious that the new assignments were its death blow. Shifting the hydrogen atoms to their keto locations made the size differences between the purines and pyrimidines even more important than would be the case if the enol forms existed. Only by the most special pleading could I imagine the polynucleotide backbone bending enough to accommodate irregular base sequences. Even this possibility vanished when Francis came in. He immediately realized that a like-with-like structure would give a 34 Å crystallographic repeat only if each chain had a complete rotation every 68 Å. But this would mean that the rotation angle between successive bases would be only 18 degrees, a value Francis believed was absolutely ruled out by his recent fiddling with the models. Also Francis did not like the fact that the structure gave no explanation for the Chargaff rules (adenine equals thymine, guanine equals cytosine). I, however, maintained my lukewarm response to Chargaff's data. So I welcomed the arrival of lunchtime, when Francis' cheerful prattle temporarily shifted my thoughts to why undergraduates could not satisfy *au pair* girls.

After lunch I was not anxious to return to work, for I was afraid that in trying to fit the keto forms into some new scheme I would run into a stone wall and have to face the fact that no regular hydrogen-bonding scheme was compatible with the X-ray evidence. As long as I remained outside gazing at the crocuses, hope could be maintained that some pretty base arrangement would fall out. Fortunately, when we walked upstairs, I found that I had an excuse to put off the crucial model-building step for at least several more hours. The metal purine and pyrimidine models, needed for systematically checking all the conceivable hydrogen-bonding possibilities, had not been finished on time. At least two more days were needed before they would be in our hands. This was much too long even for me to remain in limbo, so I spent the rest of the afternoon cutting accurate representations of the bases out of stiff cardboard. But by the time they were ready I realized that the answer must be put off till the next day. After dinner I was to join a group from Pop's at the theater.

When I got to our still empty office the following morning, I quickly cleared away the papers from my desk top so that I would have a large, flat surface on which to form pairs of bases held together by hydrogen bonds. Though I initially went back to my like-with-like prejudices, I saw all too well that they led nowhere. When Jerry came in I looked up, saw that it was not Francis, and began shifting the bases in and out of various other pairing possibilities. Suddenly I became aware that an adenine-thymine pair held together by two hydrogen bonds was identical in shape to a guanine-cytosine pair held together by at least two hydrogen bonds. All the hydrogen bonds seemed to form naturally; no fudging was required to make the two types of base pairs identical in shape. Quickly I called Jerry over to ask him whether this time he had any objection to my new base pairs.

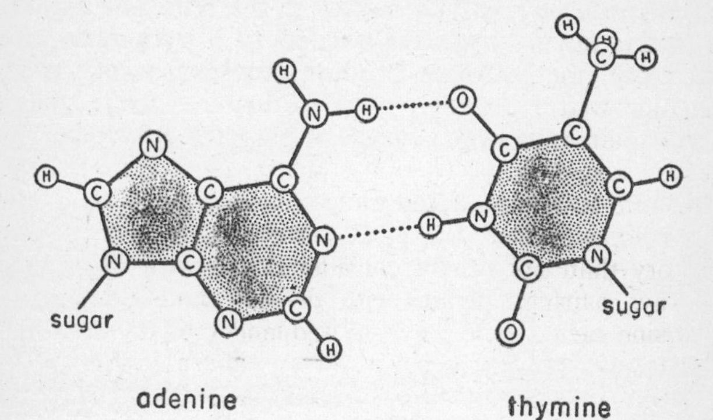

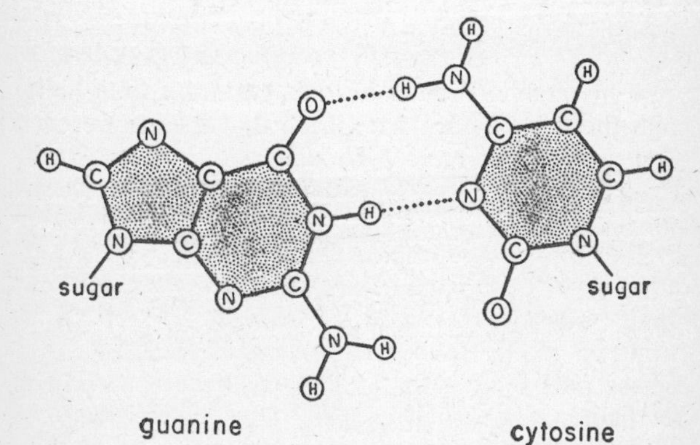

The adenine-thymine and guanine-cytosine base pairs used to construct the double helix (hydrogen bonds are dotted). The formation of a third hydrogen bond between guanine and cytosine was considered, but rejected because a crystallographic study of guanine hinted that it would be very weak. Now this conjecture is known to be wrong. Three strong hydrogen bonds can be drawn between guanine and cytosine.

When he said no, my morale skyrocketed, for I suspected that we now had the answer to the riddle of why the number of purine residues exactly equaled the number of pyrimidine residues. Two irregular sequences of bases could be regularly packed in the center of a helix if a purine always hydrogen-bonded to a pyrimidine. Furthermore, the hydrogen-bonding requirement meant that adenine would always pair with thymine, while guanine could pair only with cytosine. Chargaff's rules then suddenly stood out as a consequence of a double-helical structure for DNA. Even more exciting, this type of double helix suggested a replication scheme much more satisfactory than my briefly considered like-with-like pairing. Always pairing adenine with thymine and guanine with cytosine meant that the base sequences of the two intertwined chains were complementary to each other. Given the base sequence of one chain, that of its partner was automatically determined. Conceptually, it was thus very easy to visualize how a single chain could be the template for the synthesis of a chain with the complementary sequence.

Upon his arrival Francis did not get more than halfway through the door before I let loose that the answer to everything was in our hands. Though as a matter of principle he maintained skepticism for a few moments, the similarly shaped A-T and G-C pairs had their expected impact. His quickly pushing the bases together in a number of different ways did not reveal any other way to satisfy Chargaff's rules. A few minutes later he spotted the fact that the two glycosidic bonds (joining base and sugar) of each base pair were systematically related by a diad axis perpendicular to the helical axis. Thus, both pairs could be flipflopped over and still have their glycosidic bonds facing in the same direction. This had the important consequence that a given chain could contain both purines and pyrimidines. At the same time, it strongly suggested that the backbones of the two chains must run in opposite directions.

The question then became whether the A-T and G-C base pairs would easily fit the backbone configuration devised during the previous two weeks. At first glance this looked like a good bet, since I had left free in the center a large vacant area for the bases. However, we both knew that we would not be home until a complete model was built in which all the stereochemical contacts were satisfactory. There was also the obvious fact that the implications of its existence were far too important to risk crying wolf. Thus I felt slightly queasy when at lunch Francis winged into the Eagle to tell everyone within hearing distance that we had found the secret of life.

27

FRANCIS' preoccupation with DNA quickly became full-time. The first afternoon following the discovery that A-T and G-C base pairs had similar shapes, he went back to his thesis measurements, but his effort was ineffectual. Constantly he would pop up from his chair, worriedly look at the cardboard models, fiddle with other combinations, and then, the period of momentary uncertainty over, look satisfied and tell me how important our work was. I enjoyed Francis' words, even though they lacked the casual sense of understatement known to be the correct way to behave in Cambridge. It seemed almost unbelievable that the DNA structure was solved, that the answer was incredibly exciting, and that our names would be associated with the double helix as Pauling's was with the alpha helix.

When the Eagle opened at six, I went over with Francis to talk about what must be done in the next few days. Francis wanted no time lost in seeing whether a satisfactory three-dimensional model could be built, since the geneticists and nucleic-acid biochemists should not misuse their time and facilities any longer than necessary. They must be told the answer quickly, so that they could reorient their research upon our work. Though I was equally anxious to build the complete model, I thought more about Linus and the possibility that he might stumble upon the base pairs before we told him the answer.

That night, however, we could not firmly establish the double helix. Until the metal bases were on hand, any model building would be too sloppy to be convincing. I went back to Pop's to tell Elizabeth and Bertrand that Francis and I had probably beaten Pauling to the gate and that the answer would revolutionize biology. Both were genuinely pleased, Elizabeth with sisterly pride, Bertrand with the idea that he could report back to International Society that he had a friend who would win a

Nobel Prize. Peter's reaction was equally enthusiastic and gave no indication that he minded the possibility of his father's first real scientific defeat.

The following morning I felt marvelously alive when I awoke. On my way to the Whim I slowly walked toward the Clare Bridge, staring up at the gothic pinnacles of the King's College Chapel that stood out sharply against the spring sky. I briefly stopped and looked over at the perfect Georgian features of the recently cleaned Gibbs Building, thinking that much of our success was due to the long uneventful periods when we walked among the colleges or unobtrusively read the new books that came into Heffer's Bookstore. After contentedly poring over *The Times,* I wandered into the lab to see Francis, unquestionably early, flipping the cardboard base pairs about an imaginary line. As far as a compass and ruler could tell him, both sets of base pairs neatly fitted into the backbone configuration. As the morning wore on, Max and John successively came by to see if we still thought we had it. Each got a quick, concise lecture from Francis, during the second of which I wandered down to see if the shop could be speeded up to produce the purines and pyrimidines later that afternoon.

Only a little encouragement was needed to get the final soldering accomplished in the next couple of hours. The brightly shining metal plates were then immediately used to make a model in which for the first time all the DNA components were present. In about an hour I had arranged the atoms in positions which satisfied both the X-ray data and the laws of stereochemistry. The resulting helix was right-handed with the two chains running in opposite directions. Only one person can easily play with a model, and so Francis did not try to check my work until I backed away and said that I thought everything fitted. While one interatomic contact was slightly shorter than optimal, it was not out of line with several published values, and I was not disturbed. Another fifteen minutes' fiddling by Francis failed to find anything wrong, though for brief intervals my stomach felt uneasy when I saw him frowning. In each case he became satisfied and moved on to verify that another interatomic contact was reasonable. Everything thus looked very good when we went back to have supper with Odile.

Our dinner words fixed on how to let the big news out. Maurice, especially, must soon be told. But remembering the fiasco of sixteen months before, keeping King's in the dark made sense until exact coordinates had been obtained for all the atoms. It was all too easy to fudge a successful series of atomic contacts so that, while each looked almost acceptable, the whole collection was energetically impossible. We suspected that we had not made this error, but our judgment conceivably might be biased by the biological advantages of complementary DNA molecules. Thus the next several days were to be spent using a plumb line and a measuring stick to obtain the relative positions of all atoms in a single nucleotide. Because of the helical symmetry, the locations of the atoms in one nucleotide would automatically generate the other positions.

After coffee Odile wanted to know whether they would still have to go into exile in Brooklyn if our work was as sensational as everyone told her. Perhaps we should stay on in Cambridge to solve other problems of equal importance. I tried to reassure her, emphasizing that not all American men cut all their hair off and that there were scores of American women who did not wear short white socks on the streets. I had less success arguing that the States' greatest virtue was its wide-open spaces where people never went. Odile looked in horror at the prospect of being long without fashionably dressed people. Moreover, she could not believe that I was serious, since I had just had a tailor cut a tightly fitting blazer, unconnected with the sacks that Americans draped on their shoulders.

The next morning I again found that Francis had beaten me to the lab. He was already at work tightening the model on its support stands so that he could read off the atomic coordinates. While he moved the atoms back and forth, I sat on the top of my desk thinking about the form of the letters that I soon could write, saying that we had found something interesting. Occasionally, Francis would look disgusted when my daydreams kept me from observing that he needed my help to keep the model from collapsing as he rearranged the supporting ring stands.

By then we knew that all my previous fuss about the importance of Mg^{++} ions was misdirected. Most likely Maurice and Rosy were right in insisting that they were

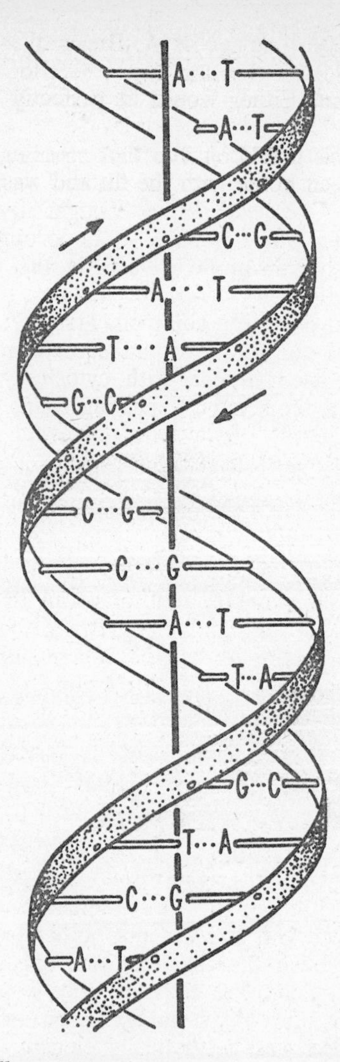

A schematic illustration of the double helix. The two sugar-phosphate backbones twist about on the outside with the flat hydrogen-bonded base pairs forming the core. Seen this way, the structure resembles a spiral staircase with the base pairs forming the steps.

looking at the Na^+ salt of DNA. But with the sugar-phosphate backbone on the outside, it did not matter which salt was present. Either would fit perfectly well into the double helix.

Bragg had his first look late that morning. For several days he had been home with the flu and was in bed when he heard that Crick and I had thought up an ingenious DNA structure which might be important to biology. During his first free moment back in the Cavendish he slipped away from his office for a direct view. Immediately he caught on to the complementary relation between the two chains and saw how an equivalence of adenine with thymine and guanine with cytosine was a logical consequence of the regular repeating shape of the sugar-phosphate backbone. As he was not aware of Chargaff's rules, I went over the experimental evidence on the relative proportions of the various bases, noticing that he was becoming increasingly excited by its potential implications for gene replication. When the question of the X-ray evidence came up, he saw why we had not yet called up the King's group. He was bothered, however, that we had not yet asked Todd's opinion. Telling Bragg that we had got the organic chemistry straight did not put him completely at ease. The chance that we were using the wrong chemical formula admittedly was small, but, since Crick talked so fast, Bragg could never be sure that he would ever slow down long enough to get the right facts. So it was arranged that as soon as we had a set of atomic coordinates, we would have Todd come over.

The final refinements of the coordinates were finished the following evening. Lacking the exact X-ray evidence, we were not confident that the configuration chosen was precisely correct. But this did not bother us, for we only wished to establish that at least one specific two-chain complementary helix was stereochemically possible. Until this was clear, the objection could be raised that, although our idea was aesthetically elegant, the shape of the sugar-phosphate backbone might not permit its existence. Happily, now we knew that this was not true, and so we had lunch, telling each other that a structure this pretty just had to exist.

With the tension now off, I went to play tennis with Bertrand, telling Francis that later in the afternoon I would write Luria and Delbrück about the double helix. It was so arranged that John Kendrew would call up Maurice to say that he should come out to see what Francis and I had just devised. Neither Francis nor I wanted the task. Earlier in the day the post had brought a note from Maurice to Francis, mentioning that he was now about to go full steam ahead on DNA and intended to place emphasis on model building.

28

MAURICE needed but a minute's look at the model to like it. He had been forewarned by John that it was a two-chain affair, held together by the A-T and G-C base pairs, and so immediately upon entering our office he studied its detailed features. That it had two, not three, chains did not bother him since he knew the evidence never seemed clear-cut. While Maurice silently stared at the metal object, Francis stood by, sometimes talking very fast about what sort of X-ray diagram the structure should produce, then becoming strangely noiseless when he perceived that Maurice's wish was to look at the double helix, not to receive a lecture in crystallographic theory which he could work out by himself. There was no questioning of the decision to put guanine and thymine in the keto form. Doing otherwise would destroy the base pairs, and he accepted Jerry Donohue's spoken argument as if it were a commonplace.

The unforeseen dividend of having Jerry share an office with Francis, Peter, and me, though obvious to all, was not spoken about. If he had not been with us in Cambridge, I might still have been pumping for a like-with-like structure. Maurice, in a lab devoid of structural chemists, did not have anyone about to tell him that all the textbook pictures were wrong. But for Jerry, only Pauling would have been likely to make the right choice and stick by its consequences.

The next scientific step was to compare seriously the experimental X-ray data with the diffraction pattern predicted by our model. Maurice went back to London, saying that he would soon measure the critical reflections. There was not a hint of bitterness in his voice, and I felt quite relieved. Until the visit I had remained apprehensive that he would look gloomy, being unhappy that we had seized part of the glory that should have gone in full to him and his younger colleagues. But there was no trace of resentment on his face, and in his subdued way he was thoroughly excited that the structure would prove of great benefit to biology.

He was back in London only two days before he rang up to say that both he and Rosy found that their X-ray data strongly supported the double helix. They were quickly writing up their results and wanted to publish simultaneously with our announcement of the base pairs. *Nature* was a place for rapid publication, since if both Bragg and Randall strongly supported the manuscripts they might be published within a month of their receipt. However, there would not be only one paper from King's. Rosy and Gosling would report their results separately from Maurice and his collaborators.

Rosy's instant acceptance of our model at first amazed me. I had feared that her sharp, stubborn mind, caught in her self-made antihelical trap, might dig up irrelevant results that would foster uncertainty about the correctness of the double helix. Nonetheless, like almost everyone else, she saw the appeal of the base pairs and accepted the fact that the structure was too pretty not to be true. Moreover, even before she learned of our proposal, the X-ray evidence had been forcing her more than she cared to admit toward a helical structure. The positioning of the backbone on the outside of the molecule was demanded by her evidence and, given the necessity to hydrogen-bond the bases together, the uniqueness of the A-T and G-C pairs was a fact she saw no reason to argue about.

At the same time, her fierce annoyance with Francis and me collapsed. Initially we were hesitant to discuss the double helix with her, fearing the testiness of our previous encounters. But Francis noticed her changed attitude when he was in London to talk with Maurice about details of the X-ray pictures. Thinking that Rosy wanted nothing to do with him, he spoke largely to Maurice, until he slowly perceived that Rosy wanted his crystallographic advice and was prepared to exchange unconcealed hostility for conversation between equals. With obvious pleasure Rosy showed Francis her data, and for the first time he was able to see how foolproof was her assertion that the sugar-phosphate backbone was on the outside of the molecule. Her past uncompromising statements on this

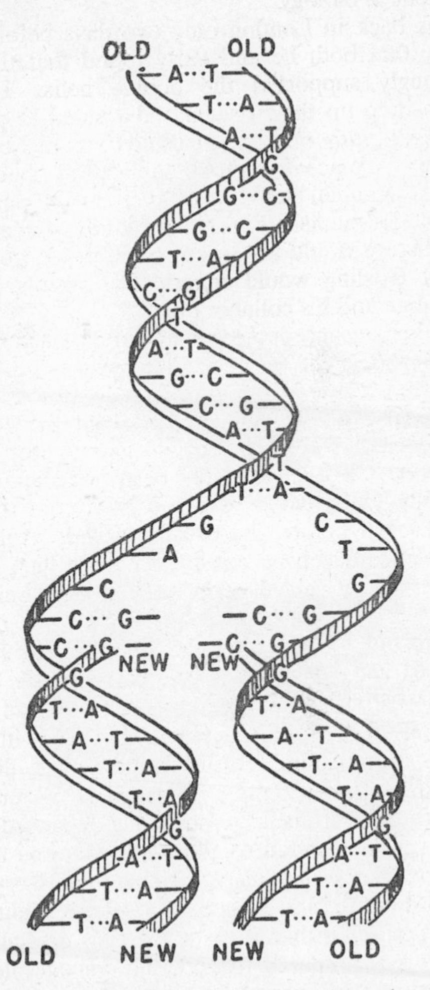

The manner envisaged for DNA replication, given the complementary nature of the base sequences in the two chains.

matter thus reflected first-rate science, not the outpourings of a misguided feminist.

Obviously affecting Rosy's transformation was her appreciation that our past hooting about model building represented a serious approach to science, not the easy resort of slackers who wanted to avoid the hard work necessitated by an honest scientific career. It also became apparent to us that Rosy's difficulties with Maurice and Randall were connected with her understandable need for being equal to the people she worked with. Soon after her entry into the King's lab, she had rebelled against its hierarchical character, taking offense because her first-rate crystallographic ability was not given formal recognition.

Two letters from Pasadena that week brought the news that Pauling was still way off base. The first came from Delbrück, saying that Linus had just given a seminar during which he described a modification of his DNA structure. Most uncharacteristically, the manuscript he had sent to Cambridge had been published before his collaborator, R. B. Corey, could accurately measure the interatomic distances. When this was finally done, they found several unacceptable contacts that could not be overcome by minor jiggling. Pauling's model was thus also impossible on straightforward stereochemical grounds. He hoped, however, to save the situation by a modification suggested by his colleague Verner Schomaker. In the revised form the phosphate atoms were twisted 45 degrees, thereby allowing a different group of oxygen atoms to form a hydrogen bond. After Linus' talk, Delbrück told Schomaker he was not convinced that Linus was right, for he had just received my note saying that I had a new idea for the DNA structure.

Delbrück's comments were passed on immediately to Pauling, who quickly wrote off a letter to me. The first part betrayed nervousness—it did not come to the point, but conveyed an invitation to participate in a meeting on proteins to which he had decided to add a section on nucleic acids. Then he came out and asked for the details of the beautiful new structure I had written Delbrück about. Reading his letter, I drew a deep breath, for I realized that Delbrück did not know of the complementary double helix at the time of Linus' talk. Instead, he was referring to the like-with-like idea. Fortunately, by the time my letter reached Cal Tech the base pairs had fallen out. If they had not, I would have been in the dreadful position of having to inform Delbrück and Pauling that I had impetuously written of an idea which was only twelve hours old and lived only twenty-four before it was dead.

Todd made his official visit late in the week, coming over from the chemical laboratory with several younger colleagues. Francis' quick verbal tour through the structure and its implications lost none of its zest for having been given several times each day for the past week. The pitch of his excitement was rising each day, and generally, whenever Jerry or I heard the voice of Francis shepherding in some new faces, we left our office until the new converts were let out and some traces of orderly work could resume. Todd was a different matter, for I wanted to hear him tell Bragg that we had correctly followed his advice on the chemistry of the sugar-phosphate backbone. Todd also went along with the keto configurations, saying that his organic-chemist friends had drawn enol groups for purely arbitrary reasons. Then he went off, after congratulating me and Francis for our excellent chemical work.

Soon I left Cambridge to spend a week in Paris. A trip to Paris to be with Boris and Harriet Ephrussi had been arranged some weeks earlier. Since the main part of our work seemed finished, I saw no reason to postpone a visit which now had the bonus of letting me be the first to tell Ephrussi's and Lwoff's labs about the double helix. Francis, however, was not happy, telling me that a week was far too long to abandon work of such extreme significance. A call for seriousness, however, was not to my liking—especially when John had just shown Francis and me a letter from Chargaff in which we were mentioned. A postscript asked for information on what his scientific clowns were up to.

29

PAULING first heard about the double helix from Delbrück. At the bottom of the letter that broke the news of the complementary chains, I had asked that he not tell Linus. I was still slightly afraid something would go wrong and did not want Pauling to think about hydrogen-bonded base pairs until we had a few more days to digest our position. My request, however, was ignored. Delbrück wanted to tell everyone in his lab and knew that within hours the gossip would travel from his lab in biology to their friends working under Linus. Also, Pauling had made him promise to let him know the minute he heard from me. Then there was the even more important consideration that Delbrück hated any form of secrecy in scientific matters and did not want to keep Pauling in suspense any longer.

Pauling's reaction was one of genuine thrill, as was Delbrück's. In almost any other situation Pauling would have fought for the good points of his idea. The overwhelming biological merits of a self-complementary DNA molecule made him effectively concede the race. He did want, however, to see the evidence from King's before he considered the matter a closed book. This he hoped would be possible three weeks hence, when he would come to Brussels for a Solvay meeting on proteins in the second week of April.

That Pauling was in the know came out in a letter from Delbrück, arriving just after I returned from Paris on March 18. By then we didn't mind, for the evidence favoring the base pairs was steadily mounting. A key piece of information was picked up at the Institut Pasteur. There I ran into Gerry Wyatt, a Canadian biochemist who knew much about the base ratios of DNA. He had just analyzed the DNA from the T2, T4, and T6 group of phages. For the past two years this DNA was said to have the strange property of lacking cytosine, a feature obviously impossible for our model. But Wyatt now said that he, together with Seymour Cohen and Al Hershey, had evidence that these phages contained a modified type of cytosine called 5-hydroxy-methyl cytosine. Most important, its amount equaled the amount of guanine. This beautifully supported the double helix, since 5-hydroxy-methyl cytosine should hydrogen-bond like cytosine. Also pleasing was the great accuracy of the data, which illustrated better than any previous analytical work the equality of adenine and thymine and guanine with cytosine.

While I was away Francis had taken up the structure of the DNA molecule in the A form. Previous work in Maurice's lab had shown that crystalline A-form DNA fibers increase in length when they take up water and go over into the B form. Francis guessed that the more compact A form was achieved by tilting the base pairs, thereby decreasing the translational distance of a base pair along the fiber axis to about 2.6 Å. He thus set about building a model with tilted bases. Though this proved more difficult to fit together than the more open B structure, a satisfactory A model awaited me upon my return.

In the next week the first drafts of our *Nature* paper got handed out and two were sent down to London for comments from Maurice and Rosy. They had no real objections except for wanting us to mention that Fraser in their lab had considered hydrogen-bonded bases prior to our work. His schemes, until then unknown to us in detail, always dealt with groups of three bases, hydrogen-bonded in the middle, many of which we now knew to be in the wrong tautomeric forms. Thus his idea did not seem worth resurrecting only to be quickly buried. However, when Maurice sounded upset at our objection, we added the necessary reference. Both Rosy's and Maurice's papers covered roughly the same ground and in each case interpreted their results in terms of the base pairs. For a while Francis wanted to expand our note to write at length about the biological implications. But finally he saw the point to a short remark and composed the sentence: "It has not escaped our notice that the specific pairing we have postulated immediately suggests a possible copying mechanism for the genetic material."

Sir Lawrence was shown the paper in its nearly final form. After suggesting a minor stylistic alteration, he enthusiastically expressed his willingness to post it to *Nature* with a strong covering letter. The solution to the structure was bringing genuine happiness to Bragg. That the result came out of the Cavendish and not Pasadena was obviously a factor. More important was the unexpectedly marvelous nature of the answer, and the fact that the X-ray method he had developed forty years before was at the heart of a profound insight into the nature of life itself.

The final version was ready to be typed on the last weekend of March. Our Cavendish typist was not on hand, and the brief job was given to my sister. There was no problem persuading her to spend a Saturday afternoon this way, for we told her that she was participating in perhaps the most famous event in biology since Darwin's book. Francis and I stood over her as she typed the nine-hundred-word article that began, "We wish to suggest a structure for the salt of deoxyribose nucleic acid (DNA). This structure has novel features which are of considerable biological interest." On Tuesday the manuscript was sent up to Bragg's office and on Wednesday, April 2, went off to the editors of *Nature*.

Linus arrived in Cambridge on Friday night. On his way to Brussels for the Solvay meeting, he stopped off both to see Peter and to look at the model. Unthinkingly Peter arranged for him to stay at Pop's. Soon we found that he would have preferred a hotel. The presence of foreign girls at breakfast did not compensate for the lack of hot water in his room. Saturday morning Peter brought him into the office, where, after greeting Jerry with Cal Tech news, he set about examining the model. Though he still wanted to see the quantitative measurements of the King's lab, we supported our argument by showing him a copy of Rosy's original B photograph. All the right cards were in our hands and so, gracefully, he gave his opinion that we had the answer.

Bragg then came in to get Linus so that he could take him and Peter to his house for lunch. That night both Paulings, together with Elizabeth and me, had dinner with the Cricks at Portugal Place. Francis, perhaps because of Linus' presence, was mildly muted and let Linus be charming to my sister and Odile. Though we drank a fair amount of Burgundy, the conversation never got animated and I felt that Pauling would rather talk to me, clearly an unfinished member of the younger generation, than to Francis. The talk did not last long, since Linus, still on California time, was becoming tired, and the party was over at midnight.

Elizabeth and I flew off the following afternoon to Paris, where Peter would join us the next day. Ten days hence she was sailing to the States on her way to Japan to marry an American she had known in college. These were to be our last days together, at least in the carefree spirit that had marked our escape from the Middle West and the American culture it was so easy to be ambivalent about. Monday morning we went over to the Faubourg St. Honoré for our last look at its elegance. There, peering in at a shop full of sleek umbrellas, I realized one should be her wedding present and we quickly had it. Afterwards she searched out a friend for tea while I walked back across the Seine to our hotel near the Palis du Luxembourg. Later that night with Peter we would celebrate my birthday. But now I was alone, looking at the long-haired girls near St. Germain des Prés and knowing they were not for me. I was twenty-five and too old to be unusual.

EPILOGUE

VIRTUALLY everybody mentioned in this book is alive and intellectually active. Herman Kalckar has come to this country as professor of biochemistry at Harvard Medical School, while John Kendrew and Max Perutz both have remained in Cambridge, where they continue their X-ray work on proteins, for which they received the Nobel Prize in Chemistry in 1962. Sir Lawrence Bragg retained his enthusiastic interest in protein structure when he moved in 1954 to London to become director of the Royal Institution. Hugh Huxley, after spending several years in London, is back in Cambridge doing work on the mechanism of muscle contraction. Francis Crick, after a year in Brooklyn, returned to Cambridge to work on the nature and operation of the genetic code, a field of which he has been the acknowledged world leader for the past decade. Maurice Wilkins' work remained centered on DNA for some years until he and his collaborators established beyond any doubt that the essential features of the double helix were correct. After then making an important contribution to the structure of ribonucleic acid, he has changed the direction of his research to the organization and operation of nervous systems. Peter Pauling now lives in London, teaching chemistry at University College. His father, recently retired from active teaching at Cal Tech, at present concentrates his scientific activity both on the structure of the atomic nucleus and on theoretical structural chemistry. My sister, after being many years in the Orient, lives with her publisher husband and three children in Washington.

All of these people, should they desire, can indicate events and details they remember differently. But there is one unfortunate exception. In 1958, Rosalind Franklin died at the early age of thirty-seven. Since my initial impressions of her, both scientific and personal (as recorded in the early pages of this book), were often wrong, I want to say something here about her achievements. The X-ray work she did at King's is increasingly regarded as superb. The sorting out of the A and B forms, by itself, would have made her reputation; even better was her 1952 demonstration, using Patterson superposition methods, that the phosphate groups must be on the outside of the DNA molecule. Later, when she moved to Bernal's lab, she took up work on tobacco mosaic virus and quickly extended our qualitative ideas about helical construction into a precise quantitative picture, definitely establishing the essential helical parameters *and* locating the ribonucleic chain halfway out from the central axis.

Because I was then teaching in the States, I did not see her as often as did Francis, to whom she frequently came for advice or when she had done something very pretty, to be sure he agreed with her reasoning. By then all traces of our early bickering were forgotten, and we both came to appreciate greatly her personal honesty and generosity, realizing years too late the struggles that the intelligent woman faces to be accepted by a scientific world which often regards women as mere diversions from serious thinking. Rosalind's exemplary courage and integrity were apparent to all when, knowing she was mortally ill, she did not complain but continued working on a high level until a few weeks before her death.

SIGNET
BOOKS